EAT FOR THE PLANET COOKBOOK

EAT FOR THE PLANET COOKBOOK

75 RECIPES FROM LEADERS OF THE PLANT-BASED MOVEMENT THAT WILL HELP SAVE THE WORLD

Nil Zacharias and Gene Stone

Abrams, New York

Contents

Introduction

In 2018 we published a book called *Eat for the Planet* to show people that our current food system, dominated by industrial animal agriculture, is quite possibly the most destructive industry on the planet. Industrial animal agriculture is not only the leading driver of climate change, but it is also responsible for air and water pollution, land degradation, and deforestation and is pushing several species to the brink of extinction.

But *Eat for the Planet* was designed to be far more than a book about doom and gloom. As stunning as much of the scientific research was, we wrote the book to help people understand that we still have a fighting chance, if we can manage to kick-start an urgent global sustainable food movement.

The book presented the straightforward idea that the real battle for the future of our planet—and the future of the human race—is being fought on our plates, multiple times a day, with every food choice we make. We articulated why humanity urgently needs to shift our diet away from industrial meat, dairy, and eggs toward a more plant-centric diet and presented the reader with a simple road map of how to bring about this change (*moderate* your consumption of animal-based foods, *replace* pantry and refrigerated kitchen staples with plant-based versions, and *embrace* whole plant foods).

While the road map was a great starting point, we realized that people needed more guidance as they embarked on this new path, and so we created the book you are now holding in your hands. But we didn't want this to be just another cookbook that showcases a few great recipes. Instead, we wanted to create a resource that captures the full depth and breadth of sustainable plant-based cooking.

To achieve this goal, we sourced recipes not just from leading plant-based chefs, but also from food brands, restaurants, and influencers who are on the cutting edge of culinary experimentation and food innovation. By doing so, we knew we would not only have a valuable cookbook with wonderful recipes, but we could also showcase the extensive array of emerging talent, products, techniques, and voices that make up the growing, sustainable plant-based food movement. While this shift in how we eat often starts in our kitchens, it is being propelled forward by an entire mission-aligned ecosystem of companies and experts who have just the advice (or product) you need to make eating plant-forward convenient and delicious.

When we asked these leaders in the plant-based revolution if they would create brand-new, mouthwatering recipes to help you start your own revolution in your own kitchen, they all responded with a resounding "Yes!"

The resulting book will help you change your food habits, soothe your cravings, and forge a new identity as someone who truly cares about food and the planet!

First, we get your appetite (and motivation) going with a quick primer on why you need to eat for the planet, and then explain exactly how you can prepare your kitchen for plant-based cooking, including details on how to stock your pantry, refrigerator, and freezer, time-saving strategies, batch-cooking techniques, ingredients and equipment basics, and other tips.

We then dive into the food, featuring drool-worthy recipes for breakfast and brunch, appetizers, soups, stews, salads, and sides, wraps, burgers, tacos, pasta, and last but not least, desserts. We've also seasoned the recipe pages with critical environmental facts and infographics that serve as reminders on why shifting away from meat, dairy, eggs, and seafood and bringing plants to the center of your plate is the single greatest step you can take to have a positive impact on the planet and future generations.

We are proud to present to you more than seventy-five extraordinary recipes from people who care about food and everything that our food choices represent. We suspect that before long, you'll be putting your own unique twists on these recipes, experimenting with new spices and ingredients, reading up on some of the contributors' work, and maybe adding in a little bit of your own grandma's secret sauce. We encourage you to make the recipes your own and share them with the world. Eating plant-based is remarkably easy because the food is healthy, fun to make, and so incredibly tasty you'll wonder how you ever ate without this book. Best of all, each bite you take helps the planet—what more can you ask for in a meal?

Nearly

6 in 10

Americans
are interested in

EATING
LESS MEAT.

Why Eat for the Planet?
A Primer

Compared to the age of our planet—4.5 billion years—the 200,000 years of human existence are barely a blip. Dinosaurs roamed the Earth for approximately 170 million years, and jellyfish are still going strong after appearing some 500 million years ago. While humankind barely registers in geological time, we have already fundamentally changed the composition of our planet—and not in a good way.

The most obvious reason is population growth: It took nearly 200,000 years for the human population to reach 1 billion, but only 200 more years for it to reach 7.5 billion. In that short period, technology evolved and hand-driven methods of extracting resources from the Earth were replaced with machine-driven methods. As humans began extracting ever more resources from the environment, we had little notion of our effects on planetary ecosystems. Consider this: We are just 0.01 percent of all life on Earth, but we have been responsible for the destruction of 83 percent of wild mammals.[1]

You might think the main reason for this is the steady march of urbanization or development, but the single greatest cause of our environmental destruction is the food we eat.

Over the past two centuries, agriculture has been completely transformed, with humans turning farms into factories with one simple goal: to produce massive quantities of meat, eggs, and milk at the lowest possible cost. Global meat production has increased fivefold since 1950, in the process engorging our land with soybeans, corn, barley, and other crops meant solely for livestock.[2] With the world's population expected to reach nearly 10 billion by midcentury, today's version of animal agriculture will soon push our planet's finite resources to the brink.

This is already happening. You likely know that 71 percent of our planet is covered with water, but did you know that nearly half the land is occupied by farm animals and the crops grown to feed them? In comparison, we occupy just 10 percent of the world's land.[3] And as the human population grows, the farm animal population must grow as well to keep up with the demand for meat and dairy.

99% of meat, dairy, and eggs
available from the U.S. comes from
FACTORY FARMS.

Today more than 20 billion farm animals are on the planet—most confined to tiny cages—and are destined to be or produce food for us. Feeding these 20 billion animals is a monumental task that is stressing our resources to the breaking point. Thirty-three percent of the Earth's arable land is devoted to growing food for animals.[4] Globally, beef cattle ranching is one of the largest drivers of deforestation,[5] with 80 percent of the deforestation of the Amazon rainforest attributed to beef production.[6] We need so much land because raising animals for slaughter is incredibly inefficient.

At the cost of one acre of land, we get a yield of 250 pounds of beef—that's about 1,000 quarter-pound hamburger patties per acre. If we used that land to grow food for humans instead of animals, we could produce 50,000 pounds of tomatoes, up to 40,000 pounds of potatoes, or 20,000 pounds of apples.[7]

Plowing all this land is also devastating our coinhabitants of the planet. Thanks to rain-forest destruction, up to 137 plant, animal, and insect species are lost every day.[8] The world is currently experiencing the sixth mass extinction event in its history, involving the highest rate of species die-off since the loss of dinosaurs 65 million years ago. Since 1970, there has already been a 58 percent decline in the number of fish, mammals, birds, and reptiles worldwide, with species becoming extinct as much as a thousand times more frequently compared to the 60 million years before humans came along.[9] A recent report from the UN Intergovernmental Science-Policy Platform on Biodiversity and Ecosystem Services found nearly 1 million of the estimated 8 million plant and animal species on Earth are at risk of extinction, many within the next few decades.[10]

Animal agriculture is so inefficient that for every 100 calories of grain fed to animals, we get back only 40 calories of milk, 22 calories of eggs, 12 calories of chicken, 10 calories of pork, and 3 calories of beef. If the world's population grows to nearly 10 billion as projected, we'll need to produce more food in the next forty years than has been created in the past ten thousand years.[11] This simply cannot happen with today's food system. However, if we decided to grow food for humans instead of farm animals, we could potentially produce enough to feed our growing population.

60% OF GLOBAL BIODIVERSITY LOSS

results from meat-based diets.

There is a very good reason that the old adage goes, "Water is life." Every living organism requires water to survive. Seventy-one percent of the planet's surface may be water, but only about 2.5 percent of all water is fresh—and the vast majority of that is locked up in glaciers and snowfields. Just about 1 percent of that fresh water is actually accessible for human consumption. Of the world's 7.5 billion people, 800 million suffer from water scarcity, largely because nearly a quarter of all fresh water is devoted to livestock.[12]

But what does water have to do with meat? Almost every hamburger, chicken nugget, and slab of pork starts with the same basic (and cheap) ingredients: soy and corn. In the United States, 47 percent of soy and 60 percent of corn is consumed by livestock.[13] These crops require a great deal of water to grow: It takes 216 gallons of water to produce one pound of soy,[14] and 108 gallons of water to produce one pound of corn.[15] The average cow will consume around one thousand pounds of feed every few months until it reaches slaughter weight. All told, it takes 1,800 gallons of water on average to produce a single pound of meat.[16] To put that in perspective, the average 22,500-gallon swimming pool has just enough water to produce 12.5 pounds of meat. Now, for the same 1,800 gallons of water required for a pound of meat, you could produce 8 pounds of tofu, grow 8 pounds of avocados, or 60 pounds of potatoes.[17] Even milk and cheese use an inordinate amount of water. That's because dairy cows are fed alfalfa, which requires 114 gallons of water per pound.[18] Taking this into account, a dairy cow producing 7 gallons of milk daily ultimately requires 4,781 gallons of water *per day*.

Simply stated, if we used all this water to produce food for direct human consumption, rather than the roundabout process of industrial animal farming, we'd have a lot more food and a whole lot more water to go around. If you avoided eating all animal products for a single day and adopted a plant-based diet, your diet would require 1,100 fewer gallons of water—enough to meet the daily indoor needs of about eleven people in the United States.[19] By 2025, two-thirds of the world's population may face chronic water shortages—a problem that will only be exacerbated by massive droughts wrought by climate change.[20]

If you have been paying attention to the science of climate change, you probably know how fossil fuels such as coal, oil, and natural gas release carbon dioxide (CO_2) into the atmosphere, forming a thick blanket that traps heat and causes global temperatures to rise. At this rate, by the year 2100 scientists expect the Earth to be on average four degrees Celsius warmer than in preindustrial times, which spells disaster for plants, animals, and humans alike.[21]

And here's a dirty little secret: The livestock system is responsible for 14.5 percent of all greenhouse emissions.[22] The industry releases more greenhouse gases than all the world's transportation—that's all the automobiles, planes, trains, and ships—combined. In fact, emissions generated during the production of a typical eight-ounce steak are equivalent to driving a small car for nearly thirty miles.[23]

Producing meat and dairy is a very oily business. From crop to table, it takes massive amounts of fossil fuels to feed our appetite for animal protein. This includes everything from feed to transportation to the use of synthetic fertilizers and pesticides to the staggering amount of energy required to power factory farms. Overall, the production of one calorie of animal protein requires about ten times as much input of fossil fuel energy as is needed for one calorie of plant protein.[24]

Animal agriculture's effects on climate change don't end there. The massive amount of deforested land required to graze farm animals and grow their crops means there are fewer trees available to convert CO_2 into oxygen. Moreover, when you chop down a tree, you release all of its stored CO_2 back into the atmosphere. Meanwhile, sustaining 20 billion farm animals also means sustaining 20 billion pairs of lungs expelling CO_2 into the atmosphere. The planet's 1.5 billion cattle are also warming the Earth

with their belches and farts. During digestion, ruminants like cows naturally produce methane—a greenhouse gas with twenty-eight times the warming potential as CO_2.[25] All told, methane from livestock accounts for 37 percent of all greenhouse gas emissions from agriculture.[26]

Each cow raised by humans uses
2-5 acres of land

And there are
1.5 billion cows on Earth.

If we keep eating this way, greenhouse gas emissions from all food production will increase 80 percent by 2050, portending disaster for the planet.[27] Average temperatures around the world will rise and extreme weather events will become more frequent and more deadly. Some parts of the world will experience heat waves and drought, which will fuel intense wildfires, cause dust storms, and impact food production and water quality. We're already seeing this happen with increased instances of wildfires in California

The average world citizen needs to eat

90% LESS PORK   75% LESS BEEF   50% LESS EGGS

to AVOID A CLIMATE CATASTROPHY.

and Australia. Warming temperatures will also increase air pollution and the spread of airborne illnesses. Other parts of the world will experience severe storms and excessive rainfall, resulting in overflowing rivers and lakes, intense flooding, and damage to life and property. Rising sea levels, meanwhile, have already forced people on the Marshall Islands and Kiribati to relocate. Forecasts indicate that their entire countries will be completely submerged within five decades.

We can do better. By shifting to eating primarily composed of plant-based foods, you can cut the carbon footprint associated with your diet in half. The resulting benefit to the world would be immense. According to one study, by transitioning toward a diet that is predominantly plant-based, we could reduce global mortality by 6 to 10 percent and food-related greenhouse gas emissions by 29 to 70 percent,

and the economic benefits of the change could be $1 trillion to $31 trillion by midcentury, thanks to fewer damaging environmental impacts, premature deaths worldwide, and lost working days, as well as reduced health care costs.[28]

Business as usual is not going to cut it any longer. It's time for each and every one of us to recognize that the industrial livestock system is at the very heart of our environmental crisis. It's time to eat for the planet!

Getting Your
Kitchen Ready

When you choose to *eat* for the planet, you'll want to be able to *cook* for the planet, and that's where the recipes in this book can help. Before we dig into the recipes, however, the following pages include some useful tips for a smooth-running plant-based kitchen.

If you're new to plant-based cooking, it's important to know that it does not have to be especially time-consuming. With a well-stocked kitchen and some time-saving strategies, you'll be well on your way to eating healthy, home-cooked, plant-based meals.

Out with the old

Not everyone is going to go all plant-based, all the time. But if you do decide that plant-based is your goal, before you stock your kitchen with ingredients, you'll want to consider eliminating all animal-based products, including any food made with meat, animal fat, or meat stock. Also eliminate eggs and all dairy products, including cheese, milk, yogurt, butter, ice cream, sour cream, coffee creamers, mayonnaise, and any salad dressings or other prepared foods containing dairy or eggs.

The good news is that virtually all these animal products can be replaced by plant-based ingredients and products.

A well-stocked pantry

One key to effortless plant-based cooking is stocking your kitchen with the right ingredients. This includes the nonperishable ingredients on your pantry shelf (the canned, dried, and bottled ingredients); the stash of prepared foods that you keep in your freezer; and the fresh vegetables, fruits, and other perishables that you keep in the fridge.

Your kitchen should include a variety of vegetables, fruits, herbs, dried or canned beans, pasta, rice and other grains, basic seasonings, nondairy milk, nuts, seeds, and nut butters, as well as flours, spices, and other standard baking items. (Make sure your spices, baking powder, and flours are fresh for optimal flavor and results.)

Grocery list

What follows is a fairly comprehensive grocery list, broken down by where the ingredients are stored (pantry, freezer, refrigerator). You don't have to purchase every ingredient on the list, and you may well already have many of these items on hand. You will want to tailor the list to your likes and dislikes (and those of your family). In addition to some basic ingredients (such as plant-based milk, fresh produce, and so on), choose some recipes that you want to try and buy the ingredients to make those recipes. Do this every week, and before long, you'll have a well-stocked kitchen.

Not included in this list are general pantry items such as dried herbs and spices, flour, cornstarch, and salt.

On the pantry shelf

- artichoke hearts, canned or jarred
- beans, canned and dried: chickpeas, lentils, white beans, black beans, kidney beans, pinto beans
- chipotles in adobo sauce
- coconut milk, unsweetened (canned)
- dairy-free chocolate chips
- dried chiles
- dried fruit: dates, apricots, cranberries, raisins, figs, mangoes
- dried mushrooms

- grains: quinoa, barley, millet, oats, cornmeal, bulgur, couscous
- jackfruit, packed in water or brine
- liquid smoke
- miso paste
- nutritional yeast (which many of us now refer to simply as "nooch")
- oils: olive oil, toasted sesame oil, neutral vegetable oil
- olives
- pasta and rice noodles
- rice: brown, basmati, jasmine, Arborio
- roasted red peppers
- soy sauce: tamari, mushroom sauce (vegan oyster sauce)
- sweeteners: agave nectar, beet sugar, molasses, maple syrup
- tahini and nut butters
- tomato products (canned: diced, whole, puree, paste; sun-dried: dehydrated or oil-packed)
- vegan soup base
- vinegars: balsamic, wine, and rice

In the freezer

- breads and doughs: whole-grain breads, tortillas, pita bread, pizza dough, puff pastry, filo dough, flatbreads
- frozen fruits and vegetables (as a backup to fresh fruits and vegetables)
- ground flaxseed meal
- homemade vegetable broth
- nuts and seeds: cashews, pistachios, sesame seeds, pine nuts, walnuts, sunflower seeds, pumpkin seeds, pecan pieces, roasted peanuts
- pesto, tomato paste
- plant-based "meats," such as veggie burgers, burger crumbles, and veggie sausages
- tofu in blocks

In the refrigerator

- cooked beans or legumes
- cooked brown rice
- fresh herbs
- hummus
- plant milk (soy, oat, etc.)
- plant-based proteins: seitan, tempeh, tofu
- vegan cheese
- vegetables and fruits, including dark leafy greens, salad greens (washed and dried), and ready-to-eat, sliced fresh fruit

Time-saving strategies

Here are some ways to save time in the kitchen:

Keep a well-stocked pantry: The surest way to make certain you can get dinner on the table is to have a well-stocked pantry, so you can always be just minutes away from a healthy, great-tasting meal.

Wash and dry fresh produce when you bring it home from the market: This ensures that your ingredients are ready when you need them and allows you to remove any wilted leaves and so on. Exceptions to this strategy are mushrooms, berries, and other fragile ingredients that should be washed immediately before using.

Pre-chop vegetables and fruits and store them in airtight containers in the fridge or freezer. Freeze cherry tomatoes for easy additions to recipes. Drain and freeze tofu. Double tofu recipes and keep cooked tofu for quick snacks or additions to wraps, salads, and tacos. Double muffin and pancake recipes and freeze the extras—they are just as delicious when rewarmed a week later.

Keep your kitchen well organized: This makes it easier, when you're ready to cook, to assemble your *mise en place*, which means gathering the equipment you'll need and measuring out ingredients in advance. Organize your pantry shelves so you know where everything is at a glance.

Read and reread a recipe: When you are familiar with your recipe, and you have your ingredients and equipment at hand, you will be amazed at how much more easily you can prepare a meal. Good prep can also help avoid kitchen mishaps, such as missing ingredients or burning dinner while you search for a spoon.

Be flexible: While it's best to plan ahead and make sure in advance that you have everything you need, it sometimes happens that you run out of an ingredient at the last minute. In those cases, rather than dropping everything to rush out to the store, try to determine if you have something in the house that can be substituted. To avoid running out of the ingredients you use most frequently, keep an ongoing grocery list in the kitchen so you can write down items the minute you run out or see that you're getting low.

Plan ahead

The best way to guarantee that you can get dinner on the table quickly involves some advance planning in the form of an ongoing grocery list and a menu plan for the week. Here are some ways to plan ahead:

Keep a list of your family's favorite dishes and rotate them regularly. If you are new to plant-based cooking, you can still make your family favorites by simply swapping out animal products and using the plant-based alternatives.

Make meals in advance, serving make-ahead, one-dish meals that you just need to reheat on especially busy nights.

Plan your menus: This doesn't have to be a complete formal menu plan. Instead, just make a brief note, such as "Monday: chili, Tuesday: tempeh stir-fry, Wednesday: pasta and salad," and so on. Having an idea of your menu for the week will help you with your grocery shopping and save you time all week long. Refer to this list when you make your grocery list, so you'll have all your ingredients on hand.

Keep your grocery list handy to jot down items as they become low.

Keep a variety of condiments on hand that add flavor to recipes, such as soy sauce, Sriracha sauce, chutneys, and salsas.

Batch cooking

When time is at a premium, consider doing a weekly cooking session during which you prepare several meals at once. Set aside a few hours to spend in the kitchen and prepare a few dishes to get through the week. Make things that reheat well or that can be portioned and frozen, such as a pot of chili, a hearty soup, a casserole, or a grain pilaf. It's also a great time to cook a big batch of brown rice, dried beans, or vegetable stock to portion and freeze.

Here are some guidelines for a batch-cooking session:

Portion and freeze: Cook a large pot of a staple grain or beans, then portion and freeze them for later use. When you need them, just thaw and heat.

Double batch: Prepare double batches of long-cooking recipes, such as stews, soups, or chili. Bonus: Their flavor improves when reheated, so they're even better when served later in the week or after being frozen for a time.

Double prep: Double up on prep work, such as chopping onions, when making more than one recipe, so you have enough for both. When you need only half an onion, chop the whole onion and refrigerate the unused portion in a sealed bag.

Make homemade vegetable stock: Rinse vegetables and greens thoroughly to get rid of sand or dirt. Place discarded trimmings (stalks, root tips, tomato hearts, vegetables that are past their prime but not rotten, greens, ends of carrots, the tough inside shells of onions, etc.) into a stockpot and cover with water. Add a few bay leaves, 1 teaspoon whole peppercorns, and 2 peeled cloves of garlic. Bring to a boil over high heat, then reduce the heat to low and simmer, covered, for 20 minutes. Strain the liquid and portion it into containers, discarding the boiled trimmings and seasonings. Stock will keep in the refrigerator for a week, and indefinitely in the freezer.

Plant-based proteins

Protein-rich foods such as tofu, tempeh, and seitan are popular plant-based alternatives to meat.

Tofu is a hearty food that has the ability to absorb flavors, making it extremely versatile. Also known as bean curd, tofu is made from ground, cooked soybeans in a process similar to the way cheese is made. Tofu comes in several varieties. Silken or soft tofu blends into a smooth, creamy consistency and is good for making sauces, desserts, and smoothies. Hard or firm tofu retains its shape, is generally higher in fat, and can be sliced or crumbled. Tofu that is packed in water should be drained before using. Hard tofu can be drained and pressed to remove excess water—to do so, place two paper towels on a plate, then the tofu, then two more paper towels, then a cutting board, then a heavy object (such as a book or a bowl filled with water). Let sit for at least 30 minutes. Crumble, slice, or dice it, and add it to the skillet with your favorite vegetables and seasonings.

Try draining and freezing blocks of tofu. After it thaws out, frozen tofu soaks up marinades easily because it has been drained and then dehydrated further by the freezing process. Freezing tofu also changes its consistency slightly, making it chewier.

Marinate tofu the way you would chicken or fish—with herbs, citrus juice, cracked black pepper, vinegar, tamari, and/or wine. Cook marinated tofu under the broiler in a skillet sprayed with cooking oil, or on a sprayed grill, until it is nicely browned.

Tempeh is made from fermented, compressed soybeans and is well suited to using in stews, stir-fries, and sautés. Tempeh turns a crisp golden brown when fried and it marinates well. Originating in Indonesia, tempeh is high in protein with a chewy texture. Tempeh can be found in the refrigerated or freezer sections of natural foods stores, Asian markets, and some supermarkets, and is usually sold in 8-ounce (225-g) slabs. The slabs can be cut into strips, cubed, or grated. Tempeh requires refrigeration, where it will keep, unopened, for several weeks. Once opened, it should be wrapped tightly and used within three days. Tempeh will keep for a month or so frozen. As tempeh can have a strong nutty flavor, it is often suggested to steam tempeh for 20 minutes before using in a recipe to mellow the flavor and make it more digestible.

Seitan is a wonderful alternative to chicken or beef because it slices and dices easily without falling apart. It is made from wheat in a process that extracts the gluten, or wheat protein. Seitan is an extremely versatile ingredient owing to its chewy texture and the forms and flavors it can take on. It can be diced, cut into strips for stir-fries, cubed for stews and soups, shredded or ground, stuffed like a roast chicken, thinly sliced, or made into sausage, loaves, and burgers.

Ready-made foods

There was a time—not many years ago—when if you wanted plant-based alternatives to animal foods, you'd have to make them yourself. These days, however, plant-based food products are common in small-town grocery stores, and new plant-based foods come out practically every month. Meat- and dairy-free consciousness has become so prevalent that restaurants that once offered an occasional vegan option on their menus now provide a page of vegan choices, and this goes for everything from the sandwich shop to high-end restaurants. We are also seeing the emergence of plant-based "butcher" shops, where top-quality, organic, non-GMO vegetables are prepared as prime cuts.

Most well-stocked supermarkets offer rows of plant-based milks made from cashews, oats, and soy, sitting right next to cow's milk in the dairy case, just as vegan buttery spreads perch alongside dairy butter. Also available are numerous vegan varieties of ice cream, cheese, yogurt, sour cream, and cream cheese. Even chain supermarkets and big-box stores carry meat alternatives such as plant-based sausage, burgers, cold cuts, bacon, and meatballs, as well as vegan seafood.

For reasons related to health, ethics, and, of course, the environment, food is better when it's made with plant-based ingredients rather than animal products. Plant-based food is better still when it's homemade and not processed. However, it is often impractical to make all your ingredients from scratch. That is where plant-based food products can help. When there's no time to cook, it's far better for you (and the planet) to quickly heat and serve ready-made plant-based products than to eat animal products. Whether you regularly eat ready-made plant-based foods such as burgers and sausages or prefer to enjoy them as occasional treats, the sheer quantity and quality of products now available is simply amazing. You'll notice that several of the recipes in this book call for such products, giving you the opportunity to try some of these foods for yourself.

If beef was swapped for beans **worldwide, we would be**

75% CLOSER

to meeting our greenhouse gas reduction goals **by 2020.**

Kitchen equipment

Like any kitchen, a plant-based kitchen requires all the basics for efficient food preparation.

Baking dishes and pans: The most useful sizes are a 9 by 13-inch (23 by 33-cm) baking dish, an 8-inch (20-cm) square baking dish, a variety of casserole dishes, rimmed baking sheets, and cake and pie pans.

Blender: A high-powered blender (such as a Vitamix or Blendtec) can be a wise investment if you do a lot of cooking at home. The blender is reserved for smoothies, sauces, soups, and anything else you want to make super-smooth and creamy very quickly. Other useful blenders are the immersion (stick) blender, which is handy for pureeing soups and sauces right in the pot, as well as a personal-size blender, which is great for small amounts of liquids, such as salad dressings and sauces.

Food processor: A food processor is essential for making pesto, pureeing vegetables, chopping nuts, and making breadcrumbs. It is also great for making pie dough, chopping vegetables, and numerous other mixing and chopping tasks. The trick is knowing when it will be faster to cut, whisk, or chop by hand, and that can usually be determined by the quantity of food involved. In addition to a large-capacity processor, some people also have a smaller model that they use for smaller tasks.

Knives: You can accomplish any task with just three knives: a paring knife for peeling and trimming; a long serrated knife for slicing bread, tomatoes, and other fragile foods; and a good 8- or 10-inch (20- or 25-cm) chef's knife for virtually everything else. Buy the best quality knives you can afford, and keep them sharp. You can chop more quickly and safely with sharp knives than dull ones.

Microwave: Ideal when you need a small amount of melted vegan butter, chocolate, or hot liquid, a microwave can also be used to soften hard winter squashes to make them easier to cut.

Other kitchen tools: You'll also need the various items used to prepare foods, including a cutting board, mixing bowls, baking pans, measuring cups and spoons, whisks, spatulas, a vegetable peeler, and a colander. It's also handy to have a box grater, salad spinner, a hand mixer or stand mixer, parchment paper, and cheesecloth.

Pots and saucepans: A 5- to 6-quart (4.7- to 5.7-L) pot or Dutch oven with a lid can be used for soups and stews or as a pasta pot. A 1-quart (960-ml) and 2-quart (2-L) saucepan (with a steamer insert) can take care of the rest.

Rice cooker: Great for making perfect rice every time.

Skillets: A good 12-inch (30.5-cm) skillet can handle most needs. It can be used to sauté, stir-fry, and braise. A good nonstick skillet is indispensable. A cast-iron skillet is great for certain things, such as cornbread, but it's heavy for everyday use. A smaller skillet is handy when you have only a small amount of ingredients to cook.

Storage containers: Keep a variety on hand for batch cooking (freezer-to-oven casserole dishes are especially handy).

Toaster oven: Perfect for toasting bread and also for baking sweet potatoes or foil-wrapped beets when you don't want to heat a large oven.

More kitchen tips

Before moving on to the recipes, here are a few more tips and tricks to use in the kitchen.

Cutting, slicing, and chopping: For maximum nutrition, leave the skins on your vegetables. Most skins are edible and delicious, if they are simply scrubbed first to remove dirt, wax, and pesticides. Chop and slice vegetables into different shapes for a more interesting presentation and varied taste. For example, beets can be sliced into thin or thick rounds, diced into cubes, or shredded raw for salads and garnish. Smaller shapes cook more quickly. For example, take care not to cook shredded carrots for as long as you would carrots sliced into rounds.

Roasting peppers and fresh chiles: Toast fresh chiles and peppers over a gas flame, on the grill, or in a hot skillet until the skins are browned and puffing away from the flesh. Place toasted chiles in a covered container or closed paper bag for 10 minutes, then peel the skins away from the flesh. Cut the peppers in half and remove the seeds and veins. Slice into strips. Make extra and freeze for future use.

Storing greens: Lettuce and leafy greens will keep much longer if you take a few minutes to prepare them for the refrigerator. Fill the sink with cold water and separate the leaves from the head. Remove any brown bits. Allow the greens to soak for a few minutes or longer to rehydrate them. Layer them in dish towels or paper towels and place them in the hydrator.

Storing other foods: Guacamole, avocado slices, and peeled or sliced fruits will keep their beautiful color if you store them in airtight containers or zip-top baggies with a splash of lemon or lime juice. Be sure to keep the mother seed with leftover avocado, or throw it in with the guacamole, to help prevent the avocado from turning an ugly shade of green.

And one more suggestion . . .

Buy local and in season

Produce that is in season and locally grown is better for you and has more flavor. Try to buy organic when you can afford to, and support your regional community by buying locally produced food from farmers' markets and food co-ops whenever possible.

In ONE MONTH of eating PLANT-BASED, ONE PERSON could SAVE

46,000 GALLONS OF WATER

13,395 POUNDS OF GRAIN

930 SQUARE FEET OF FOREST

629 POUNDS OF CO_2

Chapter One

BREAKFAST AND BRUNCH

Easy Breakfast Scramble

RECIPE BY JUST

This delicious scramble can be on the table in minutes. The recipe calls for onion, bell pepper, tomatoes, and spinach, but you can add any other mix-ins you prefer, such as sautéed zucchini or chopped vegan bacon or ham.

SERVES 2

1 teaspoon olive oil

1 tablespoon chopped onion

1 tablespoon chopped bell pepper

½ cup (120 ml) Just Egg or other vegan egg alternative

1 tablespoon diced tomato

1 small handful baby spinach

1 teaspoon minced fresh chives or green onions, plus more for garnish

Black salt (optional, see Note)

Ground black pepper

In a nonstick skillet, heat the oil over medium-high heat. Add the onion and bell pepper and cook for 1 to 2 minutes, or until onion is slightly translucent. Lower the heat to medium.

Pour the egg into the skillet with the onion and pepper. Begin to cook the egg, stirring with a stiff spatula. When about halfway cooked through, add the tomato, spinach, and 1 teaspoon chives. Stir and continue cooking until the egg is set, but not too dry, using the spatula to break up the scramble into fluffy, bite-size pieces. Remove from the heat.

Sprinkle the cooked scramble with black salt for a more eggy flavor, if desired. Season with pepper and garnish with more chives. Serve immediately.

Note: Black salt, or *kala namak*, is a sulfurous salt that imparts an amazing eggy flavor. It's fairly easily found at specialty grocery stores or online.

Breakfast Tacos

RECIPE BY JUST

Breakfast tacos are a delicious way to start the day—and make an especially fun brunch when you've got overnight guests. Set up a taco-building station on a large tray and let everyone assemble their own.

MAKES 6 TACOS

2 cups (190 g) shredded red cabbage

2 medium avocados, pitted, peeled, and sliced or cubed

¼ cup (10 g) roughly chopped fresh cilantro

1 cup (140 g) crumbled vegan chorizo sausage

1 cup (240 ml) Just Egg or other vegan egg alternative

Salt and ground black pepper

6 (6-inch /15-cm) corn or flour tortillas

¾ cup (180 ml) chunky salsa or pico de gallo

Hot sauce (optional)

Place the cabbage, avocados, and cilantro in separate bowls, or arrange on a tray for your taco-building station.

In a nonstick skillet, cook the chorizo as directed, then set aside on a plate lined with paper towels. In the same skillet, scramble the egg over medium-low heat until just cooked through, about 2 minutes, breaking it up with a stiff spatula into fluffy, bite-size pieces. Season with salt and pepper, if desired. Add the scramble to the lineup of toppings.

Warm the tortillas individually in a pan over low heat for 1 to 2 minutes per side, or arrange them in a stack, wrap them in aluminum foil, and bake in a preheated 350°F (175°C) oven for 10 to 15 minutes, or until heated. Fill the tortillas with the desired amounts of each topping. Garnish with salsa and hot sauce, if desired. Serve immediately.

MUNG BEAN SPROUTS

Mung beans, which grow on vines in Asia, are legumes (pulses) that are an Earth-friendly source of protein. Its sprouts are used to make the egg substitute Just Egg, which is lighter on the Earth than hen's eggs. The latter mostly come from high-intensity factory farms, notorious for the air and water pollution generated by chicken waste. Research has found that producing just one dozen eggs emits 2.7 kilograms of greenhouse gases, and one of the biggest impacts comes from all the land, fertilizer, and other inputs of growing chicken feed.

Mung beans, on the other hand, offer quality protein, can grow in dry conditions (requiring less water), and can "fix" nitrogen in the soil, maintaining soil nutrients and requiring less fertilizer. Its crop residues can be used as "green manure," plowed back into the soil to fertilize other plants. Best of all, you don't need to get the sprouts from the store: You can sprout the beans over and over at home in a jar. You will then have fresh sprouts to add to your salads and wraps.[29]

Best Breakfast Burritos

RECIPE BY DIANA MENDOZA FOR PETA

Soft tofu creates a nice eggy texture (more so than firm tofu does) and vegan cream cheese adds richness. Loaded with flavor from green chiles, salsa, and soy chorizo, these may be the best breakfast burritos you will ever cook, or taste.

SERVES 2

8 ounces (225 g) soft tofu, drained

1 russet potato, diced

¼ teaspoon salt

1 teaspoon ground turmeric

½ teaspoon chili powder

1 (4-ounce / 115-g) can hot or mild diced green chiles, drained

½ cup (70 g) crumbled vegan chorizo sausage

½ cup (120 ml) salsa

1 Roma tomato, diced

3 tablespoons chopped fresh cilantro

Hot sauce

2 (12-inch / 30.5-cm) flour tortillas

⅓ cup (75 g) vegan cream cheese

In a bowl, crumble the tofu and set aside. Pour 1 inch (2.5 cm) water into a pot fitted with a steamer basket. Bring to a boil over high heat. Reduce the heat to medium, then place the potato in the steamer basket. Cover and steam for 5 minutes, or until fork-tender. Drain and set aside.

Heat a skillet over medium heat. Combine the tofu and salt and cook for 3 to 4 minutes, stirring gently. Add the turmeric, chili powder, chiles, chorizo, and potato and cook until heated through, about 1 minute. Add the salsa, tomato, cilantro, and hot sauce to taste and cook for 1 minute. Keep warm.

Heat a large skillet over low heat. Warm the tortillas individually for 1 to 2 minutes to soften. Transfer the tortillas to a flat work surface. Spread a thin line of cream cheese down the center of each tortilla, to about 1 inch (2.5 cm) from the edges.

Arrange the scramble on top of the cream cheese on each tortilla. Fold in the top and bottom of each tortilla and roll from one side to seal up the burrito. Place one burrito, seam-side down, in the skillet for 1 to 2 minutes over medium heat, or until it becomes golden. Rotate the burrito until it is golden all over. Repeat with the second burrito. Serve hot.

Staffordshire Oatcakes

RECIPE BY CARYN HARTGLASS/GARY DE MATTEI

These hearty oatcakes are best made in advance and then reheated later. The inside tends to be a little gooey right after cooking, but they firm up perfectly when reheated in a 350°F (175°C) oven for 5 to 10 minutes. They also make wonderful English muffins that can be fork-split and toasted. Says Caryn, "We love them with a homemade apricot butter and a savory spread made with tahini, miso, and nutritional yeast. It's sweet and salty. Gary says it's like a jam-and-cheese sandwich."

SERVES 4

1½ cups (360 ml) warm water

1½ cups (360 ml) warm unsweetened plant milk

1 tablespoon beet sugar (optional)

1 (.25-ounce / 7-g) package active dry yeast

2½ cups (220 g) old-fashioned oats

2 cups (255 g) all-purpose gluten-free flour, such as Bob's Red Mill Gluten-Free 1-to-1 Baking Flour

½ to 1 teaspoon salt (optional)

Canola oil

In a large bowl, combine the water and milk. Stir in the sugar, if using, and the yeast, and let stand for 5 minutes, or until frothy (see Note).

Using a blender, grind the oats into a flour. Pour the oats into the yeast mixture along with the flour and salt, if using. Stir well to mix thoroughly. Cover the bowl with a clean towel and let stand in a warm place for 1 hour.

Grease a griddle pan with oil and heat over medium-high heat. Spoon about ⅓ cup (75 ml) of batter onto the pan to make a thick 3- to 4-inch (7.5- to 10-cm) pancake. Cook for 2 to 3 minutes, or until bubbles rise to the surface. Flip the oatcake over and cook until slightly browned. Repeat with the remaining batter.

Place the oatcakes on a heatproof plate in a warm oven (200°F / 90°C) until ready to serve. To store, cool to room temperature, then cover and refrigerate. Properly stored, the oatcakes will keep well in the refrigerator for up to 3 days.

Note: Sugar is not necessary to proof the yeast, but it helps speed things along. You can use less sugar, or use 1 tablespoon of flour instead for proofing. If the mixture doesn't foam, then the yeast is not active, and you'll have to start over.

Teff Almond Pancakes

RECIPE BY JENNÉ CLAIBORNE

Teff and almond flours bring a nutty flavor to these gluten-free pancakes. They are especially good topped with maple syrup and sliced bananas or other fresh fruit, such as berries or sliced peaches. Toasted nuts or shredded coconut also make for excellent additions.

MAKES 12 PANCAKES; SERVES 3 TO 4

1 cup (115 g) teff flour

½ cup (60 g) almond flour

½ cup (45 g) oat flour

1 teaspoon ground cinnamon

1 tablespoon coconut sugar or beet sugar

2 teaspoons baking powder

½ teaspoon salt

1 medium ripe banana

2 tablespoons flaxseed meal

1½ cups (360 ml) unsweetened plant milk, plus more if needed

1 teaspoon vanilla extract

Vegetable oil or vegan butter

Maple syrup, shredded coconut, toasted nuts, sliced banana, berries, or other fresh fruit, for serving

In a large bowl, combine the teff flour, almond flour, oat flour, cinnamon, sugar, baking powder, and salt. Stir well to combine.

In a separate small bowl, mash the banana. Add the flaxseed meal, 1½ cups (360 ml) milk, and the vanilla and stir well.

Pour the wet ingredients into the dry ingredients, and use a wire whisk to mix well. Add more milk if you prefer thinner pancakes.

Grease a large skillet with oil or vegan butter. Heat the skillet over medium heat. Spoon ¼ to ⅓ cup (60 to 75 ml) of batter into the skillet to make each pancake. Cook for about 4 minutes, or until bubbles rise to the surface of the pancakes and the bottoms are golden brown.

Flip the pancakes. Cook for another 3 to 4 minutes, or until the bottoms are golden. Transfer to a platter and keep warm in a 200°F (90°C) oven while you finish cooking the pancakes with the remaining batter.

Serve topped with maple syrup, shredded coconut, toasted nuts, sliced banana, berries, or other fresh fruit.

Foolproof Waffles

RECIPE BY EMILY LAVIERI-SCULL

"As much as I love making waffles or pancakes with lots of mix-ins, or whole wheat, or gluten-free," says Emily, "it's nice to have a recipe for classic, fluffy, white-flour waffles. My common alterations to this recipe are to replace the apple cider vinegar with lemon juice and lemon zest or to add some spices like cardamom and cinnamon. But most of the time, I just stick to the basic recipe and pair it with some fresh fruit and maple syrup."

SERVES 4

1 cup (240 ml) unsweetened plant milk

1 tablespoon cider vinegar

1½ cups (190 g) all-purpose flour

1 tablespoon baking powder

3 tablespoons beet sugar

½ teaspoon salt

1 teaspoon vanilla extract

3 tablespoons vegetable or canola oil

Maple syrup and fresh fruit, such as berries, sliced figs, and/or sliced peaches, for serving

Heat a waffle iron according to the manufacturer's directions.

In a small bowl, combine the milk and vinegar and allow to curdle. In a large bowl, combine the flour, baking powder, sugar, and salt.

Add the vanilla and oil to the milk mixture and stir. Pour the wet ingredients into the dry ingredients and stir until just combined.

When the waffle iron is heated, coat with nonstick cooking spray and add about ⅓ cup (75 ml) of batter. Cook for 12 to 15 minutes, or until the waffle is browned and crispy on the outside. Transfer the waffle to a cooling rack and repeat with the remaining batter. Top the waffles with maple syrup and fresh fruit and serve.

FIGS

Interested in finding an eco-friendly snack food when you don't feel like whipping up one of these recipes? Think fig. Sweet, chewy-yet-crunchy figs are good for a quick, delicious, sugary bite. Figs have been grown worldwide since biblical times and might even be the first food cultivated for the human diet. There are about 900 species growing on trees, shrubs, and vines around the world, but there's no need for imports (and transportation pollution) when so many figs are grown in California. Actually an inside-out flower, the fig feeds more than 1,300 mammal and bird species.

Figs have a mutually beneficial relationship with tiny insects called fig wasps, which pollinate the plant. The fig, in turn, helps the wasps reproduce. A wasp lays her eggs inside the unripe fruit; her female offspring then tunnel out, fly away, and continue the cycle. (Sadly, the males die inside and are absorbed back into the fruit.)

Fig trees are also resilient—they grow most anywhere, even under semi-drought conditions, so planting them can help reforesting efforts.[30]

Fruity Corn Waffles

RECIPE BY CARYN HARTGLASS/GARY DE MATTEI

Made with corn flour and studded with raisins and apples, these waffles provide the taste and aroma of a crisp autumn day. Says Caryn, "These waffles are naturally sweet, no refined sugar necessary! They have a light, cake-like texture. We love to make a double batch and freeze a bunch to be reheated in the oven or toasted on another day."

SERVES 4

2 tablespoons ground flaxseeds

2 tablespoons Ener-G Egg Replacer or other vegan egg alternative

2 tablespoons cider vinegar

1½ cups (125 g) corn flour

1 cup (125 g) all-purpose gluten-free flour, such as Bob's Red Mill Gluten-Free 1-to-1 Baking Flour

1 tablespoon baking powder

1 medium apple, cored and finely chopped

½ cup (75 g) raisins

Maple syrup, for serving

Grease and preheat a waffle iron according to the manufacturer's directions.

In a small bowl, beat the flaxseeds with ½ cup (120 ml) water. Beat in the egg replacer and vinegar.

In a large bowl, combine the corn flour, all-purpose flour, and baking powder.

Pour the flaxseed mixture and 1 cup (240 ml) water into the bowl with the flour mixture. Stir well. Add in an additional 1 cup (240 ml) water to make a thin batter.

Stir in the apple and raisins. Let the batter set for 5 to 10 minutes before pouring about ⅓ cup (75 ml) batter into the waffle iron. Cook according to the recommendations on your waffle iron, for about 10 minutes, or until the waffle is browned and crispy on the outside. Serve hot with maple syrup. Individually wrapped and tightly covered, these waffles will keep well in the refrigerator for 3 days and in the freezer for up to a month.

Wicked Baked S'mores Oatmeal

RECIPE BY CHAD SARNO/DEREK SARNO

"Who doesn't love s'mores?" say Chad and Derek. "We've paired them here with oats for those who would like to incorporate camping into their everyday breakfast routine."

SERVES 2

1 cup (90 g) old-fashioned oats

½ teaspoon ground cinnamon

1 tablespoon vanilla extract

¾ cup (38 g) vegan marshmallows

10 to 12 fresh raspberries, strawberries, and/or blueberries

¼ cup (35 g) shaved dark chocolate

4 vegan graham crackers, split into quarters

Preheat the oven to 350°F (175°C). In a small bowl, combine the oats, 2 cups (480 ml) water, the cinnamon, and vanilla and stir well. Evenly distribute the oat mixture between two small cast-iron skillets and bake the oatmeal for 15 minutes, stirring a couple of times during baking. It's done when it's wicked hot and the consistency is not too watery and not too stiff.

Remove from the oven, top with vegan marshmallows and berries, and bake for 3 more minutes to slightly melt the marshmallows.

Remove from the oven, top with shaved chocolate and graham crackers arranged nicely, and serve.

OATS

Whole-grain, heart-healthy oats are good for the environment on the farm and beyond. They're a key plant in crop rotation, in which oats are planted alternating with other crops, helping control plant diseases and insects (thus reducing pesticide use). They need less nitrogen to grow than many crops, meaning less polluting fertilizer is required. Fields of densely packed oat plants help prevent soil erosion and create shade to keep weeds from growing nearby, resulting in less herbicide use. All of this means less water pollution from chemical runoff.

Oats are even a source of renewable energy: In the Midwest, their discarded hulls are used to produce electricity and steam, reducing the need for fossil fuels. A study at the University of Iowa found that when half the coal at the campus's power plant was replaced with oat hulls, carbon dioxide emissions were reduced by 40 percent and several air pollutants were reduced by as much as 90 percent.[31]

Biscuits with Hollandaise

RECIPE BY FORA FOODS

Basic but never boring, these classic flaky biscuits are loaded with old-fashioned buttery flavor. For a tender crumb and maximum amount of flake, make sure to minimally work the dough and keep the FabaButter (a nondairy butter made with aquafaba) and almond milk chilled until ready to use.

MAKES 8 TO 10 BISCUITS

¾ cup (180 ml) plus 1 tablespoon unsweetened almond milk, divided

2 teaspoons cider vinegar

2½ cups (320 g) all-purpose flour

1 tablespoon baking powder

1 teaspoon sea salt

½ cup (115 g) cold Faba-Butter or other vegan butter

1 cup (250 ml) Hollandaise Sauce (recipe follows)

Preheat the oven to 425°F (220°C) and line a baking sheet with parchment paper.

In a glass measuring cup, whisk together ¾ cup (180 ml) milk and the vinegar with a wire whisk, and keep chilled until ready to use. In a large bowl, combine the flour, baking powder, and salt.

Cut the butter into small chunks and add to the flour. Use two butter knives or a pastry cutter to cut the butter into the flour to create a crumbly mixture with large crumbs.

Create a well in the center of the flour mixture and add the milk mixture. Using a rubber spatula, stir just until the mixture forms a ball. Transfer the dough to a clean work surface and gently knead two to four times to create a soft dough.

Pat the dough into a circle just under 2 inches (5 cm) thick. Use a floured biscuit cutter or small glass about 2½ to 3 inches (6 to 7.5 cm) in diameter to cut the dough into 8 to 10 biscuits. Gather the remaining dough and shape it into a round biscuit. Transfer to the baking sheet, use 1 tablespoon milk to brush the tops of the biscuits, and bake for 18 to 20 minutes, or until golden.

If desired, the biscuits can be split in half and grilled on a cast-iron pan. To serve, top each of the hot biscuits with about 2 tablespoons of the hollandaise.

Hollandaise Sauce

**MAKES 2 CUPS
(500 ML)**

3 ounces (85 g) raw
cashews, soaked at least
1 hour or overnight

2 tablespoons nutritional
yeast

1 teaspoon dry mustard

½ teaspoon onion powder

¼ teaspoon ground
turmeric

½ teaspoon sea salt, plus
more if needed

4 teaspoons fresh lemon
juice, plus more if needed

2 tablespoons FabaButter
or other vegan butter

½ teaspoon dried thyme
or dried tarragon, or 2
tablespoons minced fresh
parsley, cilantro, basil, or
tarragon (optional)

In a blender, pulse the cashews with 1¼ cups (300 ml) water, the
yeast, dry mustard, onion powder, turmeric, ½ teaspoon salt, and
4 teaspoons lemon juice. Blend until perfectly smooth.

In a large saucepan, melt the butter over medium-high heat. Pour
in the cashew mixture (it will sizzle!) and add the herbs, if using.
Simmer, whisking occasionally with a wire whisk, for 6 to 8 minutes.

The sauce is ready when thickened and evenly hot. Taste and adjust
the seasonings, adding a splash more lemon juice or salt if desired.
For a thinner texture, add 1 tablespoon water, or more, until the
desired consistency is reached.

Buckwheat Crêpes

RECIPE BY CARYN HARTGLASS/GARY DE MATTEI

These crêpes are ideal for brunch or a light supper. For savory crêpes, fill with grilled vegetables, mushrooms, or ratatouille. For sweet crêpes, top with berries or other fruit and sweet cream sauces made from nuts or plant milk.

SERVES 4

1⅓ cups (230 g) buckwheat groats

¼ teaspoon baking powder

Pinch of salt

Canola oil

In a high-powered blender, grind the buckwheat into a flour. Add the baking powder and salt and pulse several times to incorporate.

Add 1 cup (240 ml) water and blend, stopping periodically to scrape down the sides. Add about ¾ cup (180 ml) water, ¼ cup (60 ml) at a time, blending the mixture after each addition, until you have a thick, pourable batter. Let the batter sit for at least 15 minutes.

Lightly grease a crêpe pan with oil and wipe off the excess. Heat the pan over medium heat. After the batter has rested and the pan is hot, use a ladle to pour about ⅓ cup (75 ml) of batter into the pan. Using the back of the ladle, lightly move the batter, using a circular motion, from the center outward, making a thin, big circle. Cook for 1 minute, or until the batter appears dry on the top. Using a narrow metal spatula, gently lift up the edges of the crêpe and flip. Cook for another 30 seconds. Set the finished crêpe on a plate.

Repeat with the remaining batter. Depending on the pan you are using, it may be necessary to lightly oil the surface again to prevent sticking.

Eating a plant-based diet can decrease your environmental impact by 42-84%.

Chapter Two

APPETIZERS

Baked Artichoke Bites

RECIPE BY ROBIN ROBERTSON

"These tasty bites can be assembled quickly and make a great appetizer. The recipe is easily doubled to feed a crowd," says Robin. Serve alone or with Avocado Aioli (page 85), Hollandaise Sauce (page 41), or Spicy Avocado Crema (page 52).

MAKES 16 BITES

1 (12-ounce / 340-g) jar marinated artichoke hearts, drained, patted dry, and finely chopped

¾ teaspoon Old Bay seasoning

½ teaspoon onion powder

¼ teaspoon garlic powder

⅛ teaspoon celery salt

1 tablespoon cornstarch

1 teaspoon nori flakes or dulse flakes

1 teaspoon spicy brown mustard

½ cup (40 g) panko breadcrumbs, plus more if needed

Preheat the oven to 400°F (205°C) and line a baking sheet with parchment paper.

In a food processor, combine the artichokes, Old Bay, onion powder, garlic powder, celery salt, cornstarch, nori, and mustard. Pulse briefly to combine. Add ½ cup (40 g) breadcrumbs and pulse until well combined, leaving some texture.

Divide the mixture into 16 portions and shape into balls. If the mixture is too loose to hold together, add more breadcrumbs. Arrange the balls on the baking sheet and bake until lightly browned, about 30 minutes, turning once about halfway through. Serve hot.

Raw Sushi Rolls with Shiitake Mushrooms

RECIPE BY LENA KSANTI

Jicama stands in for the rice in these raw sushi rolls filled with shiitakes, cucumbers, avocado, red cabbage, and sprouts. A sushi rolling mat is a helpful tool for creating a good-looking roll.

MAKES 3 ROLLS

1 cup (70 g) shiitake mushrooms, stems removed and caps sliced

1½ tablespoons tamari

1 medium jicama, peeled and roughly chopped

2 teaspoons rice vinegar

3 sheets nori

2 cucumbers, cut into thin strips

1 avocado, pitted, peeled, and sliced

⅛ head red cabbage, sliced

1 cup (85 g) sunflower sprouts

In a small bowl, combine the mushrooms and tamari and allow to marinate for 30 minutes.

Meanwhile, place the jicama in a blender and pulse until it becomes small pieces about the size of rice. Place the jicama rice in a nut bag and strain out all the excess water. Transfer the jicama to a bowl and add the vinegar.

Place a sheet of nori on a bamboo rolling mat with the shiny side down. Spread a thin layer of the jicama rice onto the bottom half of the sheet.

Drain the mushrooms, reserving the soaking water, and pat dry. Layer one-third of the cucumber, avocado, mushrooms, cabbage, and sprouts on top of the jicama rice. Use the mat to create a compact roll. Dip your finger or a brush into the tamari used for marinating the mushrooms and run it along the edge of the nori to seal the roll. Use a sharp knife to cut the roll into 8 pieces. Repeat twice more with the remaining ingredients.

Hedgehog Mushroom Satay

RECIPE BY PEGGY CHAN

Hedgehog, or lion's mane, mushrooms are the star of these flavorful grilled satays served with a fragrant, luscious sauce redolent of ginger, lemongrass, and tamarind. Look for galangal, lemongrass, kaffir lime leaves, and tamarind paste in Asian markets.

MAKES 20 SKEWERS

3 tablespoons sunflower-seed oil

¼ cup (25 g) sliced, peeled fresh ginger

1 cup (75 g) crushed, chopped lemongrass

¼ cup (40 g) sliced fresh galangal

3 kaffir lime leaves

1 tablespoon tamarind paste

1 small red chile, chopped

½ cup (95 g) coconut sugar

1 teaspoon five-spice powder

1 teaspoon red pepper flakes

1¼ cups (150 g) cashews, toasted and chopped

2 cups (480 ml) coconut milk

Salt and ground black pepper

2 pounds (910 g) hedgehog mushrooms (lion's mane mushrooms), cut into bite-size pieces

In a deep skillet, heat the oil over medium-high heat. Add the ginger, lemongrass, galangal, and lime leaves and cook, stirring, until golden brown and fragrant, about 1 minute. Add the tamarind, chile, and sugar. Stir until the sugar has dissolved and is evenly incorporated, about 2 minutes. Add the five-spice powder, red pepper flakes, cashews, and coconut milk, reduce the heat to medium-low, and simmer for 15 minutes, stirring frequently. Season with salt and pepper.

Remove large pieces of herbs and aromatics from the sauce. Reduce the heat to low and allow the sauce to thicken while you prepare the satay. When the sauce reaches the desired consistency, turn off the heat, cover, and keep warm.

Soak 20 bamboo skewers in water for 10 minutes. Season the mushrooms with salt and pepper and place 4 to 5 pieces onto each skewer. Preheat the grill according to the manufacturer's instructions, then grill on medium-high heat for 5 to 8 minutes on all sides, or until grill marks appear. Remove the satay skewers from the grill and drench them in satay sauce before plating. Serve with more satay sauce on the side.

Note: You can also prepare this in your oven if you don't have a grill. Set the oven rack about 4 inches (10 cm) from the heat. Preheat the broiler. Arrange the satay skewers on a broiler pan or rimmed baking sheet. Broil for about 4 to 5 minutes, checking frequently, turning skewers once.

SUNFLOWER SEEDS AND OIL

Bees are vital to the existence of our food. Not just honey—some 70 percent of the world's food crops rely on bees for pollination; the bees help fertilize flowering plants, so the plants can make seeds and reproduce. But the bee population has dropped drastically in the last decade or so, in part due to climate change, habitat loss, and pesticides called neoconids. We can help save bees by planting bee-friendly crops—such as blueberry bushes, pumpkins, herbs such as thyme, and sunflowers—in our gardens.

Sunflowers also have green cred of their own: They're all naturally bred (no GMOs), and they're raised mostly no-till, meaning the soil they're grown in isn't turned, which helps it to store carbon. Leftover sunflower stalks help feed wildlife in winter, and sunflower fields provide bird habitat during the growing season. Seventy-eight bird species use sunflower fields as a fall migration stopover.[32]

Loaded Brussels Sprouts Nachos

RECIPE BY CHAD SARNO/DEREK SARNO

"These loaded nachos are made wicked healthy with the base of roasted Brussels sprouts," say Chad and Derek. Serve alone or with tortilla chips. The Wicked Healthy Cheeze Sauce used in this recipe is also the perfect creamy sauce to spread over pizza, use in a mac and cheese, top a sandwich, or enjoy as a decadent dip.

SERVES 4 TO 6

1 pound (455 g) Brussels sprouts, trimmed and halved

1 tablespoon olive oil

Salt and ground black pepper

2 to 3 cups (480 to 720 ml) Wicked Healthy Cheeze Sauce (recipe follows)

2 Field Roast Mexican Chipotle Sausages or other vegan sausage links, crumbled

1 cup (240 ml) fresh tomato salsa or pico de gallo

½ medium red onion, diced

½ medium avocado, pitted, peeled, and diced

½ bunch fresh cilantro, chopped and divided

Tortilla chips, for serving (optional)

Preheat the oven to 400°F (205°C). In a large cast-iron skillet, combine the Brussels sprouts, oil, and salt and pepper to taste and toss together. Roast for about 30 minutes, stirring occasionally to cook evenly, until the Brussels sprouts are tender and golden brown on the outside.

Remove from the oven and cover with the cheeze sauce, sausages, salsa, onion, avocado, half the cilantro, and a handful of tortilla chips, if using. Bake for 8 to 10 minutes, or until the Brussels sprouts are tender on the inside and golden brown on the outside.

To serve, garnish with the remaining cilantro and serve with more tortilla chips, if desired.

Wicked Healthy Cheeze Sauce

**MAKES 2 CUPS
(480 ML)**

½ large head cauliflower
(about 1¼ pounds /
570 g), roughly chopped

½ cup (60 g) raw
cashews, soaked at least
1 hour or overnight

7 cloves garlic, peeled

1 bay leaf

2 tablespoons nutritional
yeast

1 tablespoon miso paste

2 teaspoons smoked
paprika

½ teaspoon coarse sea
salt

Pinch ground black
pepper

2 (7-ounce / 200-g) pack-
ages Follow Your Heart
Pepper Jack Slices or
other vegan cheese

1 cup (240 ml) unsweet-
ened plant milk, plus
more if needed

In a small saucepan, combine the cauliflower, cashews, garlic, and bay leaf. Add enough water to cover the ingredients and bring the mixture to a boil over high heat. Reduce the heat to medium-low and simmer for 8 to 10 minutes, or until the cauliflower is soft and easy to pierce with a fork. Remove from the heat.

Pour the pot contents through a strainer, reserving ½ cup (120 ml) of the cooking liquid. Remove the bay leaf, then transfer the contents to a high-powered blender, along with the reserved cooking liquid, the yeast, miso, paprika, salt, and pepper.

Start the blender on low and slowly increase the speed to high. Use the tamper to push all the ingredients into the blade. With the blender running, separate the cheese slices and add to the mixture. As everything comes together, it will form a thick mass.

Slowly add the milk to the blender, tamping down and keeping a close eye on the consistency. When the texture is smooth, creamy, and pourable, the sauce is done. Store in a tightly covered container in the refrigerator for up to 5 days.

Spicy Avocado Crema

RECIPE BY MILKADAMIA

This spicy sauce is great served with enchiladas or tacos, or tossed with a corn salad. It also makes a great dipping sauce for Baked Artichoke Bites (page 46).

**MAKES ABOUT
¾ CUP (180 ML)**

½ medium avocado, pitted and peeled

2 tablespoons unsweetened Milkadamia or other macadamia milk

2 tablespoons fresh lime juice

½ teaspoon salt

½ teaspoon ground black pepper

½ cup (20 g) coarsely chopped fresh cilantro

2 teaspoons minced jalapeño chile

1 small clove garlic, minced

In a food processor, combine the avocado, milk, lime juice, salt, pepper, cilantro, jalapeño, and garlic. Pulse until smooth and evenly combined. Transfer to a bowl and use immediately or cover and refrigerate until needed. This sauce is best used on the same day that it is made.

Cambodian Mushroom Dip

RECIPE BY ROBIN ROBERTSON

"My husband developed an intense craving for a sublime dip he'd tried at a restaurant in Siem Reap," says Robin, "so I developed this version for him that hits all the right flavor notes." If oyster mushrooms are unavailable, substitute white button mushrooms. Omit the chile (or add more) depending on your heat preference.

**MAKES ABOUT
2 CUPS (480 ML)**

1 tablespoon olive oil

¼ cup (35 g) finely chopped shallot

1¼ cups (90 g) finely chopped oyster mushrooms

1 clove garlic, finely chopped

½ small red chile, seeds removed and finely chopped (optional)

2 teaspoons minced lemongrass

1 teaspoon grated, peeled fresh ginger

1 teaspoon light brown sugar

1½ tablespoons ground turmeric

1 teaspoon ground coriander

2 tablespoons creamy peanut butter

1 (14-ounce / 420-ml) can coconut milk

1 tablespoon fresh lime juice

1 teaspoon soy sauce

2 tablespoons crushed roasted peanuts

2 tablespoons chopped fresh cilantro

Toasted baguette slices or raw vegetables, for serving

In a skillet, heat the oil over medium heat. Add the shallot, mushrooms, and garlic and cook, stirring, for 2 to 3 minutes to soften. Add the chile, if using, the lemongrass, ginger, brown sugar, turmeric, and coriander. Reduce the heat to medium-low and cook, stirring, for 1 minute.

Add the peanut butter, coconut milk, lime juice, and soy sauce. Cook, stirring, for 3 to 4 minutes to thicken. Taste and adjust the seasonings, if needed.

For a smoother consistency, transfer the mixture to a blender or food processor and blend until fairly smooth, retaining a little texture. For a chunkier consistency, leave as-is.

Transfer to a bowl and serve warm, garnished with the peanuts and cilantro, along with toasted baguette slices or sliced raw vegetables for dipping.

Szechuan Dumplings

RECIPE BY CHAD SARNO/DEREK SARNO

"Making dumplings is fun and easy with a little practice," say Chad and Derek. "Definitely don't be intimidated by the procedure or you'll be missing out on the meditative process of folding and assembling these dumplings." Try stuffing them with Beyond Meat Chicken, but once you get the hang of it, you'll find that dumplings can be filled with almost anything. Serve with a simple dipping sauce made with soy sauce, rice vinegar, and scallions.

MAKES 30 DUMPLINGS

2 (9-ounce / 255-g) packages Beyond Meat Chicken Strips or other vegan chicken strips

3 tablespoons toasted sesame oil, divided, plus more if needed

¼ cup (15 g) minced green onion

1 tablespoon sambal oelek (chile paste), or 1 teaspoon finely chopped fresh hot chile

3 cloves garlic, finely chopped

2½ tablespoons minced, peeled fresh ginger

3 tablespoons maple syrup

1½ cups (140 g) finely shredded green cabbage

½ teaspoon sea salt

½ teaspoon ground black pepper

1 to 2 tablespoons cornstarch

30 round vegan dumpling wrappers

In a food processor, pulse the Beyond Meat Chicken until coarsely minced. Transfer to a bowl and set aside.

In a sauté pan, heat 2 tablespoons sesame oil over medium-high heat. Add the green onion, sambal oelek, garlic, ginger, and maple syrup. Sauté for 2 to 3 minutes. Add the cabbage and cook, stirring frequently to ensure the spices caramelize but do not burn, for about 4 minutes. Remove from the heat and add to the bowl with the Beyond Meat Chicken. Season with the salt and pepper and stir to combine.

Lightly sprinkle the cornstarch over a baking sheet (this will keep your dumplings from sticking to the sheet) and fill a small bowl with water.

To prepare the dumplings, place about 1 tablespoon of dumpling filling in the center of a dumpling wrapper. Using a small bit of water on your finger, moisten around the edge of the whole wrapper. Pick up the dumpling and fold it into a taco shape. Starting with one corner, crimp around the edge, only crimping the part of the wrapper facing you. Place the dumpling on the baking sheet. Repeat with the remaining wrappers and filling.

In a sauté pan, heat the remaining 1 tablespoon sesame oil over medium heat and sear the dumplings in batches until golden on the bottom, about 3 minutes. Add more sesame oil to the pan between batches, if needed. Meanwhile, in a bamboo steamer basket or other steamer, bring water to a boil. Place the seared dumplings in the steamer. Steam for about 3 minutes. To check for doneness, insert a toothpick into the middle of a dumpling for a few seconds, then remove the toothpick and press it against your wrist. If the toothpick feels hot, the dumplings are done. Remove from the heat and serve immediately.

Macadamia Pesto Crostini

RECIPE BY MILKADAMIA

In addition to being delicious as a crostini topping, this macadamia pesto is terrific tossed with cooked pasta or vegetables.

SERVES 4 TO 6

1 cup (30 g) basil leaves

½ cup (10 g) arugula

½ cup (15 g) parsley leaves

½ cup (120 ml) Mac Nut macadamia oil

1 tablespoon fresh lemon juice

½ teaspoon red pepper flakes

2 tablespoons macadamia nuts

1 small clove garlic, peeled

1 baguette, cut into ½-inch (12-mm) slices and toasted

Olive oil, for storing, if needed

In a food processor, combine the basil, arugula, parsley, oil, lemon juice, red pepper flakes, nuts, and garlic. Pulse until combined. Transfer to a small bowl and serve with the toasted baguette slices. To store the pesto, pour a thin layer of olive oil on top of the pesto, then keep in a tightly sealed container for up to 5 days in the refrigerator or up to a month in the freezer.

If Americans shifted their diets to add more plant-based foods, the reduction of greenhouse gas emissions in ONE DAY would be equivalent to ELIMINATING 661 MILLION CAR MILES.

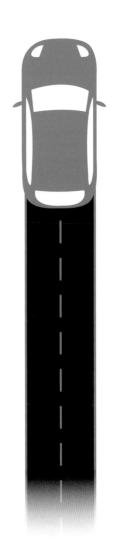

Chapter Three

SOUPS, STEWS, SALADS, AND SIDES

Potato-Leek Soup with Tarragon and Fennel Seeds

RECIPE BY ALLISON MCLAUGHLIN

"This recipe is perfect for a cold day and when you want something quickly," says Allison. Made with just a few simple ingredients, this soup is nutritious and a lighter take on traditional creamy potato-leek soup. The fennel seeds add a unique taste that complements the sweetness of the leeks perfectly.

MAKES ABOUT 8 CUPS (2 L); SERVES 4

1 teaspoon olive oil

1 small yellow onion, finely chopped

4 cloves garlic, finely chopped

4 carrots, diced

10 Yukon gold potatoes, diced

2 leeks, trimmed, thinly sliced, and rinsed to remove sand

1 teaspoon dried tarragon

2 bay leaves

½ teaspoon fennel seeds

Salt and ground black pepper

4 cups (960 ml) vegetable stock, plus more if needed

Chopped chives or scallions, for garnish

In a soup pot, heat the olive oil over medium-low heat. Add the onion and cook, stirring, until golden, about 5 minutes. Add the garlic and cook, stirring, until toasty and aromatic, about 1 minute. Add the carrots, potatoes, and leeks and stir to combine. Add the tarragon, bay leaves, and fennel seeds. Season with salt and pepper. Continue to cook for about 5 minutes, stirring, to allow the vegetables to soften slightly and become infused with the herbs.

Gently pour in 4 cups (960 ml) stock and turn up the heat to medium-high. Bring the soup to a boil, cover, reduce the heat to low, and simmer for 20 minutes. Check the potatoes and carrots to make sure they are fully cooked but not overdone. Add more stock if you prefer a thinner consistency. Remove the bay leaves and serve hot (see Note), garnished with chives or scallions. Leftovers can be refrigerated for up to 5 days or frozen for up to 1 month.

Note: If you prefer a creamier texture, use an immersion blender or a high-powered blender to blend part of the soup.

"No Chicken" Enchilada Soup

RECIPE BY NO EVIL FOODS

Need to get dinner on the table quickly? Raid your pantry and get your soup on. This is straight-up warm, comforting, cold-weather goodness and shows up with all the flavors of traditional enchiladas but without all the fussy assembly.

MAKES ABOUT 9 CUPS (2.1 L); SERVES 4

2 tablespoons olive oil or coconut oil

1 medium yellow onion, chopped

1 (9-ounce / 255-g) package No Evil Foods "No Chicken" or other vegan chicken strips

1 (4-ounce / 115-g) can diced green chiles

2 cloves garlic, finely chopped

1¼ cups (300 ml) or 1 (10-ounce / 280-g) can red enchilada sauce

1 (15-ounce / 430-g) can black beans, rinsed and drained

1 (14-ounce / 400-g) can fire-roasted diced tomatoes, with juice

1 (15-ounce / 430-g) can whole-kernel corn, drained

1 teaspoon ground cumin

1 teaspoon salt

2 cups (480 ml) vegetable stock

Chopped fresh cilantro, sliced avocado, fresh lime juice, and tortilla chips, for serving

In a large pot, heat the oil over medium-high heat until glistening.

Add the onion and cook for 3 minutes, then add the chicken and stir to combine. Cook for 3 to 5 minutes, or until the ingredients begin to brown. Add the chiles and garlic and cook, stirring, for 30 seconds, or until aromatic. Add the enchilada sauce, beans, tomatoes, corn, cumin, salt, and stock and bring to a boil, then reduce the heat to medium-low and simmer for 20 to 30 minutes.

Serve hot with cilantro, avocado, lime juice, and tortilla chips. Store the soup (ungarnished) in a tightly covered container in the refrigerator for up to 5 days or in the freezer for up to a month.

Easy Red Lentil Soup

RECIPE BY JESSICA MURNANE

"Red lentils cook up quickly, making this a good choice when you want a pot of soup in a hurry," says Jessica. "It's also easy to make a double batch and store some in the freezer. If you prefer more heat, use the entire jalapeño."

MAKES ABOUT 8 CUPS (2 L); SERVES 4 TO 6

1 tablespoon olive oil or coconut oil

1 medium onion, chopped

3 cloves garlic, finely chopped

2 carrots, chopped

2 stalks celery, chopped

1-inch (2.5-cm) piece fresh ginger, peeled and grated

1 cup (180 g) diced tomatoes

½ jalapeño chile, seeded and chopped

1 teaspoon ground cumin

1 teaspoon ground turmeric

Sea salt

4 cups (960 ml) vegetable broth or water

1 cup (190 g) red lentils, rinsed

2 teaspoons tamari or coconut aminos

Ground black pepper

In a large pot, heat the oil over medium heat. Add the onion and sauté for 5 to 7 minutes, or until soft and translucent. Add the garlic, carrots, celery, ginger, tomatoes, jalapeño, cumin, turmeric, and a pinch of salt and cook, stirring, until the vegetables become soft, about minutes 5 minutes longer. Add the broth, lentils, and tamari and stir. Bring the mixture to a boil and then reduce the heat to low.

Cover and simmer the soup for 30 minutes, stirring frequently to help break down the lentils.

Season with salt and pepper. Serve hot. Leftovers can be refrigerated for up to 5 days or frozen for up to 1 month.

Butternut Velvet Soup

RECIPE BY VEGGIE GRILL

This soup is vibrantly colored and bursting with beta carotene. It's equally suited for an autumn soup supper with warm crusty bread or as a first course at Thanksgiving dinner.

**MAKES ABOUT
10 CUPS (2.4 ML);
SERVES 4 TO 6**

1½ pounds (680 g) butternut squash, peeled, seeded, and cut into 1-inch (2.5-cm) pieces

½ pound (225 g) carrots, cut into 1-inch (2.5-cm) pieces

1 pound (455 g) red yams, peeled and cut into 1-inch (2.5-cm) pieces

½ teaspoon sea salt, divided, plus more if needed

¼ teaspoon ground black pepper

2 tablespoons vegetable oil

¼ cup (55 g) Earth Balance or other vegan butter

½ cup (55 g) chopped white onion

2 cloves garlic, chopped

1 tablespoon minced, peeled fresh ginger

¼ stalk lemongrass, finely chopped

¼ teaspoon ground cinnamon

1 teaspoon dried sage

½ cup (90 g) finely diced Granny Smith apple

5 cups (1.2 L) vegetable broth

½ cup (120 ml) unsweetened almond milk

Preheat the oven to 400°F (205°C).

Place the squash, carrots, yams, ¼ teaspoon of the salt, the pepper, and oil in a large bowl. Toss to combine, then spread evenly on a baking sheet. Bake for 10 minutes, or until slightly softened and lightly browned. Remove from the oven and set aside.

In a large soup pot, melt the butter over medium-high heat. Add the onion, garlic, ginger, and lemongrass. Cook for 5 minutes, or until the onion starts to soften and there is a little color on the garlic.

Add the cinnamon, sage, and apple. Season with the remaining ¼ teaspoon salt. Stir well. Add the roasted vegetables and the vegetable broth. Bring to a boil, reduce the heat to medium-low, and simmer for 30 minutes, or until the squash is tender. Taste and add more salt, if desired. Remove from the heat, transfer the soup to a blender, and blend until smooth, with the center part of the cover removed to let steam escape, or blend in the pot with an immersion blender. Add the milk. Stir and serve warm. Leftovers can be refrigerated for up to 5 days or frozen for up to 1 month.

Mushroom, Bean, and Barley Soup

RECIPE BY DEVORAH BOWEN

This is a delicious and hearty soup, perfect for lunch or dinner. It's packed with fiber, high in protein, and made without any oil. Try to use organic ingredients whenever possible.

MAKES ABOUT 10 CUPS (2.4 ML); SERVES 4 TO 6

½ small onion, chopped

3 cloves garlic, finely chopped

2 stalks celery, chopped

¼ teaspoon concentrated vegetable base

8 ounces (225 g) white mushrooms, sliced

¾ cup (150 g) pearled barley

1 (15-ounce / 430-g) can kidney beans, drained and rinsed

1½ teaspoons dried thyme

1½ teaspoons dried dill

1 bay leaf

Salt and ground black pepper

¼ cup (30 g) raw cashews, soaked at least 1 hour or overnight, soaking water reserved

In a large pot, over medium heat, combine the onion, garlic, celery, ¼ cup (60 ml) water, and the vegetable base. Sauté until the onion is translucent, about 5 minutes. Stir in the mushrooms and sauté until browned, about 4 minutes, pouring in more water as needed so the vegetables do not stick.

Add the barley and stir well, pouring in more water as needed. Add the beans, thyme, dill, bay leaf, and salt and pepper.

Pour in 7 cups (1.7 L) water, reduce the heat to medium-low, and simmer, covered, for about 30 minutes.

In a blender, puree the cashews with their soaking water until creamy and add to the soup. Continue to simmer until the barley is tender, about 25 minutes. Serve hot. Leftovers can be refrigerated for up to 5 days or frozen for up to 1 month.

BEANS

Kidney, butter, black, garbanzo . . . these versatile sources of quality protein, if substituted for beef, could make a big dent in fighting climate change. So says a 2017 study published in the journal *Climatic Change*, which found that if all Americans made this switch, we could achieve half to three-quarters of the greenhouse gas cuts needed to meet the 2020 targets set by former president Obama, even without changing our energy systems and other carbon-heavy activities. That's because the land and food needed to raise cattle—not to mention the methane their farts emit!—consumes so many natural resources (including chopping down forests for grazing areas and raising soybeans for feed). Beans are also pulses, which adapt well to a changing climate and take nitrogen out of the air and "fix" it into the soil.[33]

Smoky Jackfruit Stew

RECIPE BY ASHLEY MELILLO FOR THE JACKFRUIT COMPANY

This stew is brimming with good-for-you ingredients, yet you'd never know it based on taste. Its savory, rich, and oh-so-creamy base is made with a combination of nourishing cauliflower-cashew cream and vegetable broth. Smoked jackfruit is stirred in at the end to add a meaty, hearty bite to this completely plant-powered dish.

SERVES 6

4 cups (530 g) cauliflower florets

2 tablespoons olive oil

5 green onions, trimmed and thinly sliced

2 large carrots, diced

1 tablespoon smoked paprika

3 cloves garlic, finely chopped

4 cups (960 ml) low-sodium vegetable broth, divided

⅔ cup (80 g) raw cashews, soaked at least 1 hour or overnight

¼ cup (15 g) nutritional yeast

2 tablespoons fresh lemon juice

1½ teaspoons sea salt, plus more if needed

Ground black pepper

2 (10-ounce / 280-g) packages Jackfruit Company Smoked Jackfruit (see Note)

1 bunch Lacinato kale, washed, stemmed, and chopped

Bring a large pot of water to a boil over high heat. Add the cauliflower and boil for 7 to 10 minutes, or until fork-tender. Drain and set aside.

In a large pot, heat the oil over medium-low heat. Add the green onions, carrots, and paprika. Sauté for about 5 minutes, or until the vegetables begin to soften, stirring occasionally. Add the garlic and cook for 1 minute, or until it begins to soften.

Pour in 3 cups (720 ml) broth, increase the heat to medium-high, and bring to a boil. Reduce the heat to low and simmer, covered, for 10 minutes.

In a high-powered blender, puree the cauliflower with the remaining 1 cup (240 ml) broth, ½ cup (120 ml) water, the cashews, yeast, lemon juice, 1½ teaspoons salt, and pepper to taste. Blend on high speed for 2 minutes, or until completely smooth and creamy.

Add the cauliflower mixture to the soup pot and stir to incorporate. Stir in the jackfruit and kale. Increase the heat to medium and simmer for 5 to 8 minutes to thicken and reduce the stew. Generously season with more salt and pepper. Serve warm.

Leftovers can be refrigerated for up to 5 days or frozen for up to 1 month. If the soup is thicker than desired after refrigerating and reheating, simply whisk in a bit more broth or water and adjust the seasonings as needed.

Note: Pick through the jackfruit to ensure there are no tough pieces. If there are, discard them.

Loaded Zucchini and Quinoa Salad

RECIPE BY MARIA KOUTSOGIANNIS

Quinoa and lightly grilled zucchini combine with olives, prunes, and cherry tomatoes for a delicious main-dish salad. When you cook the quinoa and zucchini in advance and make the dressing ahead of time, this salad can be assembled quickly.

SERVES 4

2 large zucchini, sliced lengthwise ¼ inch (6 mm) thick

Salt and ground black pepper

2 tablespoons plus ¼ cup (60 ml) olive oil, divided

2 to 3 cups (400 to 600 g) cooked quinoa, at room temperature

½ cup (65 g) California prunes, roughly chopped

½ cup (75 g) black and green olives, roughly chopped or halved

1 cup (145 g) cherry tomatoes, halved and sprinkled with salt

¾ cup (30 g) roughly chopped fresh cilantro

¼ cup (60 ml) fresh lemon juice (about 2 large lemons)

2 tablespoons Dijon mustard

1 teaspoon dried oregano

Preheat the grill. In a bowl, toss the zucchini with salt, pepper, and 2 tablespoons oil.

Grill the zucchini slices for 2 minutes on each side, then transfer to a large bowl to cool. Once the zucchini has come to room temperature, add the quinoa, prunes, olives, cherry tomatoes, and cilantro and toss well.

In a jar, combine the lemon juice, remaining ¼ cup (60 ml) oil, the mustard, oregano, and salt and pepper. Shake well to blend the dressing. Pour the dressing over the salad, toss well, and serve.

Vacation Salad

RECIPE BY JANE ESSELSTYN

"On a family beach vacation, a friend made this salad and fell in love with it—perhaps part of the magic came from the fact that he was with his family, on vacation, creatively trying out new recipes," says Jane. "But what makes them still love it—and what makes it a versatile dish for everyone—is that this tasty topping can go on brown rice, mixed with quinoa, over a cooked sweet potato, or with crackers. Versatility is a virtue!"

SERVES 4

2 cups (350 g) roasted corn kernels (thawed, if using frozen)

1 cup (155 g) shelled edamame

½ cup (55 g) chopped red onion

½ cup (75 g) chopped red bell pepper

¼ cup (10 g) coarsely chopped fresh cilantro

2 tablespoons fresh lemon juice

3 teaspoons grated, peeled fresh ginger

¼ teaspoon ground black pepper

In a bowl, combine the corn, edamame, onion, bell pepper, cilantro, lemon juice, ginger, and black pepper. Stir until well combined and serve at room temperature, or cover and refrigerate until needed and serve chilled. Stored in an airtight container in the refrigerator, the salad will keep for up to 3 days.

Thai Slaw

RECIPE BY RIP ESSELSTYN

"This recipe, originally published in *Plant-Strong*, came from the Healthy You Network in Tucson, Arizona," says Rip. "We've changed it up here, using dairy-free yogurt and adding cucumber and a Thai chili-garlic sauce for a creamier slaw, so it delivers the flavors of a banh mi sandwich (if you wish, add tofu and French bread to complete the transcontinental experience—a terrific Thai salad on French bread). No cooking necessary." Use more or less chili-garlic sauce, depending on your heat preference.

SERVES 4 TO 6

½ head green cabbage, finely shredded

1 teaspoon salt

Juice of 1 lime

½ cup (120 ml) dairy-free yogurt

½ cup (120 ml) creamy peanut butter

1 tablespoon maple syrup

1 teaspoon tamari

2 teaspoons minced, peeled fresh ginger

2 teaspoons chili-garlic sauce

½ cup (55 g) shredded carrot

½ cup (75 g) thinly sliced cucumber

1 red bell pepper, seeded and thinly sliced

1 bunch fresh cilantro, chopped

¼ cup (30 g) raw peanuts, coarsely chopped

In a large bowl, combine the cabbage with the salt and lime juice and toss together. Let sit while you prepare the dressing.

In a small bowl, combine the yogurt, peanut butter, maple syrup, tamari, ginger, and chili-garlic sauce and stir.

Pour the dressing over the cabbage, then add the carrot, cucumber, bell pepper, cilantro, and peanuts. Stir to combine. Serve immediately or cover and refrigerate until needed. The slaw is at its best (crispiest) when served on the day it's made but can be enjoyed for up to 3 days when kept tightly covered in the refrigerator.

Mango Curry Bowls with Turmeric-Tahini Dressing

RECIPE BY REBBL

As beautiful as it is delicious, this nourishing bowl has a base of flavorful brown rice that's been cooked using REBBL 3 Roots Mango Spice Elixir. The rice is also seasoned with curry powder and then topped with protein-packed mung beans, steamed broccoli, sliced radishes, sauerkraut, microgreens, and a sprinkle of sesame seeds.

SERVES 4

FOR THE RICE:

1 cup (180 g) uncooked brown rice, rinsed

1 (12-ounce / 360-ml) bottle REBBL 3 Roots Mango Spice Elixir

1 teaspoon curry powder

Sea salt and ground black pepper

FOR THE DRESSING:

¼ cup (60 ml) fresh lemon juice (about 2 large lemons)

2 tablespoons extra-virgin olive oil

2 tablespoons tahini

1 large clove garlic, peeled

1 teaspoon ground turmeric

Sea salt and ground black pepper

FOR THE BOWLS:

1 cup (195 g) uncooked split mung beans, rinsed

Sea salt and ground black pepper

1 head broccoli, chopped

4 radishes, thinly sliced

1 cup (165 g) cubed mango

1 cup (150 g) sauerkraut

1 cup (35 g) microgreens

2 tablespoons sesame seeds

Lemon wedges, for serving

Cook the rice: In a medium saucepan, combine the rice, elixir, curry powder, and ½ cup (120 ml) water. Bring to a boil, uncovered, over high heat. Reduce the heat to low and simmer, covered, until the rice is tender, about 40 minutes. Remove from the heat. Season with salt and pepper.

Make the dressing: In a blender, combine the lemon juice, oil, tahini, garlic, and turmeric with 3 tablespoons water and blend until very smooth. Season with salt and pepper, then chill until ready to use.

Cook the beans: In a medium pot, combine the beans with 2 cups (480 ml) water. Bring to a boil, uncovered, over high heat. Reduce the heat to low and cook, covered, until tender, about 20 minutes. Remove from the heat. Season with salt and pepper.

Steam the broccoli: Pour 1 inch (2.5 cm) water into a pot fitted with a steamer basket. Bring to a boil over high heat. Reduce the heat to medium and place the broccoli in the basket. Cover and steam until the broccoli is bright green and just barely tender, about 5 minutes.

Assemble the bowls: Divide the rice among 4 bowls. Top with the beans, broccoli, radishes, mango, sauerkraut, microgreens, and sesame seeds. Serve with the turmeric-tahini dressing and lemon wedges.

Leftovers should be stored in an airtight container in the fridge— they're best enjoyed within 3 to 4 days. Keep the dressing in a separate container until just before serving; otherwise the rice will absorb the dressing.

BROCCOLI

Most people know this green cruciferous veggie is super-healthful thanks to its high levels of antioxidants. But did you know that sulforaphane, the natural chemical in broccoli that helps us fight cancer-causing free radicals, also helps the plant stave off insect pests? That means fewer chemical pesticides are needed to grow it. It can also help clear air pollution—in your body. A recent study at Johns Hopkins University found that eating broccoli helps the body quickly excrete the toxic air pollutants benzene and acrolein.

Broccoli is also one of the most water-thrifty crops, using just 285 cubic meters globally per ton of produce, one of the lowest amounts of all food crops. And since the United States is the world's largest producer of broccoli, it's easy to buy locally.[34]

Nepali Greens

RECIPE BY VICTORIA MORAN

"While traveling in Nepal, we met expatriate Tibetans who invited us to dinner and regaled us with these simple, super-tasty greens that are typical of the region," says Victoria.

SERVES 6

2 teaspoons neutral oil

1 teaspoon ground cumin

1 teaspoon garam masala

1 medium onion, chopped

2 pounds (910 g) mustard greens or kale, destemmed, well washed, and chopped

3 tomatoes, coarsely chopped

Salt

In a large skillet, heat the oil over medium heat. Add the cumin and garam masala and sauté for 30 seconds, stirring. Add the onion and sauté until translucent, about 5 minutes. Stir in the greens in stages, allowing them to cook down before adding more so that your skillet will accommodate everything. Add the tomatoes and season with salt. Reduce the heat to low and cook, stirring occasionally, for about 10 minutes. At the end of cooking, add a small amount of water, bring to a boil, and serve.

Farmhouse Caesar

RECIPE BY CIERRA DE GRUYTER, COFOUNDER NEXT LEVEL BURGER

In this healthy twist on the Caesar salad, the pepita-based dressing ensures an abundance of healthy fats, magnesium, zinc, fiber, and antioxidants. The baby kale is less bitter and more tender than mature kale leaves (and doesn't require massaging). "Have fun experimenting with different veggies and toppings," says Cierra. "The real star here is the dressing." To save time, instead of making the cashew parmesan, you can use a store-bought nondairy parmesan, such as the ones made by Go Veggie or Follow Your Heart.

SERVES 4

FOR THE DRESSING:

½ cup (65 g) raw pumpkin seeds

3 tablespoons fresh lemon juice

2½ tablespoons Dijon mustard

1 tablespoon garlic powder

1 tablespoon tahini

1 tablespoon minced roasted garlic, such as Christopher Ranch Roasted Garlic

½ teaspoon salt

¼ teaspoon ground black pepper

FOR THE CASHEW PARMESAN:

3 tablespoons raw cashews

1 tablespoon nutritional yeast

¼ teaspoon salt

⅛ teaspoon garlic powder

FOR THE SALAD:

1 (5-ounce / 140-g) package baby kale salad greens

½ cup (56 g) vegan bacon bits, such as Frontier Co-Op Organic Bac'Uns Vegetarian Bits

1 cup (180 g) diced tomatoes

1 cup (30 g) croutons

Make the dressing: In a high-powered blender, process the pumpkin seeds, lemon juice, mustard, 1 tablespoon garlic powder, the tahini, roasted garlic, ½ teaspoon salt, and the pepper with ½ cup (120 ml) water, scraping down the sides as necessary, until smooth. Refrigerate at least 4 hours to allow the flavor to deepen. The dressing will keep well in the refrigerator for up to 5 days in a tightly covered container.

Make the cashew parmesan: In a food processor or high-powered blender, combine the cashews, yeast, ¼ teaspoon salt, and ⅛ teaspoon garlic powder and process to fine crumbs. Store the cashew parm in a tightly covered container in the refrigerator for up to a month.

Assemble the salad: In a large bowl, toss the kale with the dressing and bacon bits. Divide the salad among 4 bowls or plates and top each salad with the tomatoes, croutons, and cashew parmesan.

Oven-Baked Zucchini Fries

RECIPE BY CARYN HARTGLASS/GARY DE MATTEI

The pumpkin-seed breading gives these baked zucchini fries their richness without using oil. Since the seeds are used in place of breadcrumbs, this recipe is gluten-free if you use wheat-free tamari, coconut aminos, or Bragg Liquid Aminos instead of soy sauce.

SERVES 4

¾ cup (100 g) raw pumpkin seeds

1 teaspoon onion powder

1 teaspoon garlic powder

1 teaspoon ground turmeric

¼ teaspoon ground black pepper

Pinch of salt

⅓ cup (75 ml) unsweetened plant milk

2 tablespoons tahini

1 tablespoon soy sauce

2 medium zucchini

Preheat the oven to 400°F (205°C). Line a baking sheet with parchment paper.

In a food processor, grind the pumpkin seeds into a meal. (Or grind the seeds in a spice grinder, in small batches.) Add the onion powder, garlic powder, turmeric, pepper, and salt and pulse to incorporate. Pour into a shallow bowl.

In a separate small bowl, beat the milk into the tahini, a few tablespoons at a time, with a fork until smooth. Beat in the soy sauce.

Cut the zucchini into strips similar to French fries, about ½ inch (12 mm) thick and 3 to 4 inches (7.5 to 10 cm) long.

Dip a piece of zucchini in the tahini sauce and then in the pumpkin seed "breading," making sure the zucchini is well coated. Place on the baking sheet. Repeat with the remaining zucchini.

Bake for 15 minutes. Turn the zucchini fries over and bake for another 15 minutes.

Serve immediately or refrigerate in an airtight container. Properly stored, they will keep for 3 to 4 days. To reheat, place the zucchini fries on a parchment-lined baking sheet and warm in a 350°F (175°C) oven for 10 minutes.

Crispy Smashed Potatoes
with Avocado Aioli

RECIPE BY AMY WEBSTER

"This has become one of my favorite go-to recipes for an easy and tasty side dish," says Amy. **"I love it because I don't spend so much time cutting potatoes. It's simple and everyone enjoys it."**

SERVES 2 TO 4

6 to 8 small to medium Yukon gold potatoes, not peeled

¼ cup (60 ml) extra-virgin olive oil, plus more for the baking sheet

1 to 2 teaspoons garlic powder

Salt and ground black pepper

½ cup (20 g) chopped fresh cilantro or parsley

1 cup (215 ml) Avocado Aioli (recipe follows)

Place the potatoes in a large pot and add enough water to cover the potatoes. Bring to a boil over high heat, then reduce the heat to medium and simmer, uncovered, for about 25 minutes, or until the potatoes are fork-tender. Drain the potatoes and allow them to cool for 5 minutes.

Preheat the oven to 450°F (230°C). Lightly grease a baking sheet (without parchment paper for crispier potatoes) and place the potatoes on the baking sheet. Using a cup or mug, press down on each potato until they are smashed. Some will break apart.

Drizzle ¼ cup (60 ml) oil over the potatoes and use a cooking brush to coat. Sprinkle with the garlic powder, and season with salt and pepper. Roast the potatoes in the oven for 30 to 40 minutes, or until crispy and golden.

To serve, top each potato with cilantro, avocado aioli, and more salt and pepper, if desired.

POTATOES

When it comes to starchy staples and sustainability, the humble spud beats out rice and pasta, according to a recent study published in the *Journal of Cleaner Production*. Researchers from Cranfield University in the UK compared the environmental footprints of the three foods, finding that growing potatoes consumes less water and emits fewer climate-changing greenhouse gases than the others. One reason: Potatoes not only grow faster, they also produce more tons of crop per hectare, meaning they use less land for a shorter period of time than, say, wheat used for spaghetti.[35]

Avocado Aioli

**MAKES ABOUT 1 CUP
(215 ML)**

1 large avocado, pitted
and peeled

1 clove garlic, peeled

1 tablespoon fresh lemon
juice

¼ cup (60 ml) vegan
mayo, such as Just Mayo
or Vegenaise

Salt and ground black
pepper

Small bunch cilantro
or parsley, coarsely
chopped (optional)

In a food processor, combine the avocado, garlic, lemon juice, mayo, salt, pepper, and cilantro, if using, and process until smooth. Store in a tightly sealed container until needed. This is best if used on the day that it is made.

Lemon-and-Maple Sriracha Sweet Potato Jojos

RECIPE BY CHAD SARNO/DEREK SARNO

Starchy, sweet, spicy, acidic, and herby! These thyme-kissed roasted sweet potato wedges are easy to make, and they have a crispy texture around the edges, while the insides are soft and tender.

SERVES 4 TO 6

1 teaspoon neutral oil

1 tablespoon maple syrup

1 tablespoon fresh lemon juice

1 tablespoon Ninja Squirrel Sriracha or other Sriracha sauce, plus more if needed

1 tablespoon chopped fresh thyme, divided

1 teaspoon ground black pepper

½ teaspoon salt

1 or 2 large sweet potatoes, not peeled, cut lengthwise into medium-size wedges

Preheat the oven to 350°F (175°C).

In a large bowl, combine the oil, maple syrup, lemon juice, Sriracha, half the thyme, the pepper, and salt and mix with a wire whisk to incorporate. Add the sweet potatoes and toss to coat evenly.

Arrange the sweet potato wedges on a baking sheet, skin side down. Roast in the oven for 30 minutes, or until fork-tender.

Garnish with the remaining thyme and more Sriracha, if desired. These taste best when served hot and eaten right away.

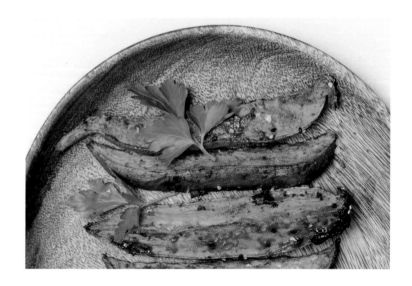

Producing one gallon of COW'S MILK releases 17.6 POUNDS OF CO$_2$ into the atmosphere.

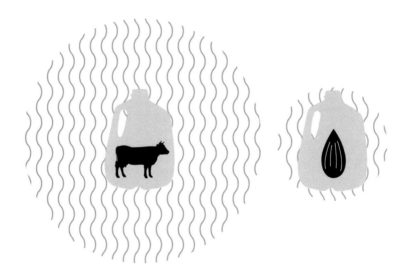

One gallon of PLANT-BASED MILK only releases around 3 POUNDS OF CO$_2$.

WRAPS, BURGERS, AND TACOS

Raw Veggie Wraps

RECIPE BY LENA KSANTI

Collard leaves are filled with crisp vegetable strips and a hearty walnut pâté with a hint of curry in these healthy wraps. Make the pâté ahead of time and cut up your veggies in advance, and these tasty wraps will come together quickly. "I especially enjoy serving the leftover pâté as a spread for crackers," says Lena.

MAKES 4 WRAPS

4 large collard leaves

½ cup (120 ml) Raw Walnut Pâté (recipe follows)

1 carrot, sliced into thin strips

2 stalks celery, sliced into thin strips

1 medium zucchini, sliced into thin strips

¼ small purple cabbage, sliced into thin strips

Cut off the protruding end of the stem of each collard leaf. Lay the leaves, top-side down, on a cutting board. Using a paring knife, gently shave down the raised part of the spines so the surface of each leaf is nice and flat.

Spread about 2 tablespoons of the pâté down the center of each leaf and distribute the carrot, celery, zucchini, and cabbage evenly among the leaves, laying the vegetables parallel to the spines, leaving about 1 inch (2.5 cm) around the edges uncovered.

Fold over the top and bottom of each leaf and then wrap up the sides, tucking the leaf tightly with every turn. Slice in half with a sharp knife. Repeat with the remaining wraps.

Raw Walnut Pâté

**MAKES ABOUT
2½ CUPS (600 ML)**

1 cup (100 g) walnuts

1 cup (140 g) chopped
carrot

1 cup (100 g) chopped
celery

4 sundried tomatoes,
soaked in boiling water
for 30 minutes, then
drained

Pinch of sea salt

Pinch of ground cumin

Pinch of curry powder

In a high-powered blender, combine the walnuts, carrot, celery, tomatoes, salt, cumin, and curry powder and blend until smooth. Transfer to a bowl, cover, and refrigerate until needed. Properly stored, the pâté will keep well in the refrigerator for up to 5 days. Tightly wrapped, it will keep in the freezer for 1 month.

Tia Blanco's Beyond Tacos

RECIPE BY TIA BLANCO FOR BEYOND MEAT

Beyond Meat burgers, the revolutionary plant-based burgers that look and cook like beef, are used to make these quick and easy tacos. If fresh corn is unavailable, use thawed frozen corn kernels instead.

SERVES 6

2 teaspoons coconut oil, divided

2¼ cups (325 g) corn kernels (cut from 3 ears of corn)

3 cloves garlic, finely chopped

4 Beyond Meat burger patties or other vegan burger patties

1 teaspoon chili powder

1 teaspoon paprika

1 teaspoon cayenne pepper

2 tablespoons vegan mayo

½ cup (55 g) chopped onion

1 jalapeño chile, seeded and minced

Juice of 1 lime

Salt and ground black pepper

6 (6-inch / 15-cm) corn or flour tortillas

½ cup (20 g) chopped fresh cilantro

6 lime wedges

In a skillet, heat 1 teaspoon oil over medium-high heat. Add the corn and cook, stirring, until slightly charred, about 3 minutes. Remove from the heat and transfer to a bowl to cool.

In the same skillet, heat the remaining 1 teaspoon oil over medium heat. Add the garlic and cook, stirring, for 1 minute, then add the burger patties. Use a spatula to break up the burgers and brown them, about 5 minutes. Add the chili powder, paprika, and cayenne. Cover, and keep warm over very low heat.

To the corn, add the mayo, onion, jalapeño, lime juice, and salt and pepper. Stir until incorporated.

Warm the tortillas individually in a pan over low heat for 1 to 2 minutes, or arrange them in a stack, wrap them in aluminum foil, and bake in a preheated 350°F (175°C) oven for 10 to 15 minutes, or until heated. Divide the burger mixture and the corn mixture evenly on top of the tortillas. Garnish with the cilantro and lime wedges.

Lentil Tacos with Roasted Cauliflower

RECIPE BY AMY WEBSTER

"I recently made this easy recipe at a family reunion where my husband and I were the only vegans out of ten people," says Amy. "As it cooked, the house filled with the scent of simmering lentils, tomato paste, and spices. The best part was that everyone loved it. My meat-eating family went back for seconds (and we enjoyed leftovers the next day, too)." These tacos can be made with potatoes instead of cauliflower, if desired.

MAKES 8 TACOS

FOR THE ROASTED CAULIFLOWER:

1 large head cauliflower (about 2½ pounds / 1.2 kg), cut into small florets

3 tablespoons olive oil

Salt and ground black pepper

FOR THE LENTIL FILLING:

1 tablespoon olive oil

¾ cup (85 g) chopped yellow or white onion

2 large cloves garlic, finely chopped

Salt

3 tablespoons tomato paste

1 teaspoon ground cumin

1 teaspoon chili powder

¾ cup (145 g) brown lentils or other lentils, rinsed and picked through

2 cups (480 ml) vegetable broth or water

TO SERVE:

8 (6-inch / 15-cm) corn or flour tortillas

1 cup (250 ml) Quick Avocado Sauce (recipe follows), other vegan aioli, or salsa

½ to 1 cup (15 to 30 g) fresh cilantro or spring greens

Make the roasted cauliflower: Preheat the oven to 425°F (220°C). In a large bowl, toss the cauliflower with 3 tablespoons oil and salt and pepper until evenly coated. Arrange the cauliflower on a baking sheet, preferably not touching and in a single layer. Bake for 30 to 35 minutes, flipping halfway, or until the cauliflower has crispy golden edges.

Make the lentil filling: In a medium pot, heat 1 tablespoon oil over medium heat. Add the onion, garlic, and salt and sauté until the onions begin to soften, about 5 minutes. Add the tomato paste, cumin, and chili powder and sauté for 1 to 2 minutes, stirring continuously. Add the lentils and broth and stir. Raise the heat and bring the mixture to a gentle simmer. Cook, uncovered, for 25 to 35 minutes, or until the lentils are tender. You may need to reduce the heat to maintain a gentle simmer. Once the lentils are done cooking, drain off any excess liquid, cover, and set aside until ready to assemble the tacos.

To serve, warm the tortillas individually in a pan over low heat for 1 to 2 minutes, or arrange them in a stack, wrap them in aluminum foil, and bake in a preheated 350°F (175°C) oven for 10 to 15 minutes, or until heated.

To assemble the tacos, place a small portion of the lentil mixture on each tortilla, add the cauliflower, drizzle with avocado sauce, and top with cilantro. Store leftovers separately in tightly sealed containers in the refrigerator for up to 3 days.

Quick Avocado Sauce

**MAKES ABOUT
2 CUPS (480 ML)**

2 medium avocados

1 tablespoon fresh lemon juice

1 tablespoon cider vinegar

Salt

A handful of fresh cilantro or parsley

In a food processor, combine the avocados, lemon juice, vinegar, salt, and cilantro and process until smooth. This sauce is best if used on the same day that it is made.

Barbecued Pulled-Mushroom Tacos

RECIPE BY JEFF STANFORD

Jeff's family's favorite foods were Mexican and barbecue, which are brought together in this recipe. Jeff's nephew even became famous for creating a pulled-pork sandwich served from a Vancouver, Canada, food truck. A disappointed Jeff wanted to show his nephew that the same flavor and texture could be produced with plants and fungi. Here barbecued shredded trumpet mushrooms are cooked to resemble pulled pork; the onions are braised to produce the texture of fat. The tacos are served with a pineapple *cruda* and pickled red onions. "The pineapple cruda makes a lot," says Jeff, "so I like to serve the leftover cruda with tortilla chips."

MAKES 8 TACOS

4 trumpet mushrooms

Himalayan pink salt or sea salt

1 medium yellow onion, thinly sliced

1 tablespoon Marinade Seasoning (recipe follows)

¼ cup (60 ml) Ravens' Barbecue Sauce (recipe follows) or other vegan barbecue sauce

8 (6-inch / 15-cm) corn or flour tortillas

¾ cup (170 g) Pineapple Cruda (recipe follows)

¾ cup (120 g) Pickled Red Onions (recipe follows)

In a food processor fitted with the grater attachment, grate the mushrooms with their stems. Process so that the mushrooms are lying lengthwise when grating. Alternatively, use a hand grater with large holes.

In a skillet or on a griddle sprayed with nonstick cooking spray, brown the grated mushrooms over medium heat for about 3 minutes, without stirring. Sprinkle lightly with salt. Turn the mushrooms using a spatula and brown the other side for 3 minutes. Transfer the mushrooms to a bowl and set aside.

In a skillet over medium heat, braise the onion in ¼ cup (60 ml) water until it is soft and translucent, about 10 minutes, adding more water if necessary so that the onion does not dry out.

Transfer the onion to the bowl with the mushrooms and toss with the marinade seasoning. Let marinate for 30 minutes at room temperature or in the refrigerator.

Preheat the oven to 350°F (175°C). When ready to serve, add the barbecue sauce to the mushroom mixture and stir to thoroughly coat. Transfer the mixture to an ovenproof skillet and heat in the oven for 4 minutes.

Warm the tortillas individually in a pan over low heat for 1 to 2 minutes, or arrange them in a stack, wrap them in aluminum foil, and bake in a preheated 350°F (175°C) oven for 10 to 15 minutes, or until heated.

Divide the mushroom mixture among the tortillas, top each taco with a heaping tablespoon of pineapple cruda, and finish by sprinkling pickled red onions over the cruda. Serve at once.

Marinade Seasoning

**MAKES ABOUT
¼ CUP (25 G)**

1 teaspoon smoked paprika

1 tablespoon brown sugar

1 teaspoon smoked salt

1 teaspoon ground cumin

1 tablespoon garlic powder

1 teaspoon cayenne pepper

1 teaspoon chili powder

In a bowl, combine the paprika, brown sugar, salt, cumin, garlic powder, cayenne, and chili powder, stir well, and store in a sealable jar or zip-top bag for up to 2 weeks.

Ravens' Barbecue Sauce

**MAKES ABOUT
2½ CUPS (600 ML)**

½ cup (55 g) chopped yellow onion

2 cloves garlic, chopped

1 (6-ounce / 170-g) can tomato paste

2 teaspoons paprika

1 teaspoon smoked paprika

2 teaspoons chili powder

1 teaspoon chipotle powder

½ cup (120 ml) cider vinegar

3 tablespoons vegan Worcestershire sauce

½ cup (120 ml) maple syrup

2 teaspoons tamari

¼ cup (60 ml) strong brewed coffee

¼ cup (60 ml) orange juice (optional)

Salt

In a heavy saucepan over medium heat, braise the onion and garlic in ¼ cup (60 ml) water until translucent, about 10 minutes, adding more water if necessary so the onion does not dry out. Stir in the tomato paste, paprika, smoked paprika, chili powder, and chipotle powder and cook for about 10 minutes to blend the flavors.

Add the vinegar, Worcestershire sauce, maple syrup, tamari, and coffee. Add the orange juice, if using, for a sweeter sauce. Reduce the heat to medium-low and cook, stirring occasionally, for about 30 minutes. Add more water if the sauce becomes too thick.

Remove from the heat and transfer to a high-powered blender. Cover, with the center part of the blender cover removed to let steam escape, and puree until smooth. If the sauce is too thin, return to the saucepan and reduce further over low heat until it reaches the desired consistency. Taste and season with salt, if necessary. Store in a sealable jar or tightly covered container in the refrigerator for up to 1 week.

ORANGES

Luscious oranges may be juicy, but like other citrus they're not thirsty in terms of their water use, at around 560 cubic meters per ton of fruit globally—versus, say, apples at more than 800 m^3 per ton and cherries at 1,600 m^3 per ton. And they do even better up against other foods such as asparagus (more than 2,000 m^3 water per ton) and almonds (more than 8,000 m^3 per ton).

Groves of orange trees suck climate-changing carbon dioxide out of the atmosphere. Research shows that each acre of mature citrus in Florida stores more than 23 tons of CO_2. The sprawling lands on which they grow offer wildlife habitat for animals, including an endangered panther. A recent study found more than 150 species have been observed visiting Florida's orange grove ecosystems.

And the juice? A 2010 study found that producing a 1-gallon carton of orange juice (from growing through squeezing, shipping, and storage) emits 7.5 pounds of CO_2. That beats 1 gallon of cow's milk, which releases more than 17 pounds—more than twice the footprint.[36]

Pineapple Cruda

**MAKES 4 TO 6 CUPS
(900 TO 1350 G)**

1 fresh ripe pineapple,
peeled and finely diced

1 medium yellow onion,
finely diced

1 shallot, minced

1 jalapeño chile, seeded
and minced

1 green bell pepper,
seeded and finely diced

½ bunch fresh cilantro,
chopped

Juice of 1 lime

In a bowl, combine the pineapple, onion, shallot, jalapeño, bell pepper, cilantro, and lime juice and let marinate for 5 to 10 minutes. Store in a tightly sealed container in the fridge for up to 4 days.

Pickled Red Onions

**MAKES ABOUT
1½ CUPS (225 G)**

1 medium red onion, thinly
sliced

¼ cup (60 ml) agave
nectar

¼ cup (60 ml) brown rice
vinegar

½ teaspoon salt

In a bowl, combine the onion, agave, vinegar, and salt and let marinate for at least 1 hour before serving. Store in a tightly sealed container in the fridge for up to 3 days.

Smoky Maple Tempeh Burgers with Chipotle Aioli

RECIPE BY TAAVI MOORE

These smoky burgers are topped with pickled red onions and a slightly spicy and creamy chipotle aioli. They are grillable, too: Simply transfer them to a hot grill after they are baked. "I recommend preparing all the components ahead of time," says Taavi, "then you can put everything out and let everyone build their own burger."

SERVES 6

FOR THE PICKLED RED ONIONS:

1 red onion, thinly sliced

2 tablespoons white vinegar

1 tablespoon beet sugar

½ teaspoon salt

1 tablespoon whole black peppercorns

FOR THE BURGERS:

8 ounces (225 g) tempeh, cut into chunks

1 tablespoon vegetable oil

½ white onion, chopped

3 cloves garlic, peeled and crushed

Salt

1 carrot, chopped

1 (15-ounce / 430-g) can black beans, rinsed and drained

4 teaspoons maple syrup

1 tablespoon tamari

1 teaspoon miso paste

½ teaspoon ground coriander

½ teaspoon paprika

1 teaspoon ground cumin

½ teaspoon liquid smoke

¼ teaspoon chili powder

Ground black pepper

Nonstick cooking spray

Flour, for dusting hands

FOR THE CHIPOTLE AIOLI:

½ cup (120 ml) vegan mayo

1 tablespoon canned chipotle chiles in adobo, chopped

2 teaspoons fresh lemon juice

½ teaspoon salt

TO SERVE:

6 burger rolls, such as ciabatta

Lettuce leaves, tomato slices, or other veggies of choice

Fresh sprouts (optional)

Make the pickled red onions: In a large mason jar, combine the onion, vinegar, sugar, ½ teaspoon salt, and the peppercorns. Add enough water to cover the onions. Cover and refrigerate for at least 30 minutes or (preferably) overnight.

Make the burgers: Preheat the oven to 375°F (190°C). Line a baking sheet with parchment paper.

Pour 1 inch (2.5 cm) water into a pot fitted with a steamer basket. Bring to a boil over high heat. Reduce the heat to medium and place the tempeh in the basket. Cover and steam for 6 minutes. Set aside.

In a skillet, heat the oil over medium heat. Add the onion and garlic and sauté for 5 to 6 minutes, or until fragrant and translucent. Season with salt. Add the carrot and cook for 5 minutes, or until tender.

Transfer the sautéed vegetables to a food processor. Add the tempeh, black beans, maple syrup, tamari, miso, coriander, paprika, cumin, liquid smoke, chili powder, and salt and pepper. Process until smooth, leaving a few chunks for texture. Add 1 tablespoon water if needed to help blend.

Lightly flour your hands and form the tempeh mixture into patties, using about ¼ cup (40 g) of the mixture for each patty. Place the patties on the baking sheet, then lightly spray them with cooking spray. Bake for 30 minutes, then flip and bake for an additional 10 minutes, or until golden brown and firm.

Make the chipotle aioli: While the burgers are baking, combine the mayo, chipotle chiles, lemon juice, and ½ teaspoon salt in a food processor and process until creamy.

To serve, slather a generous amount of aioli over a bun, then add lettuce, tomato, or other veggies of choice, a burger patty, pickled red onions, more aioli, and sprouts, if using. Repeat with the remaining burgers.

For best storage, wrap the burgers individually in foil and store them in the refrigerator for up to 3 days or in the freezer for up to 1 month.

Plant-based meats produce

10x LESS

greenhouse gas emissions

than beef.

Chapter Five

PASTA AND NOODLES

Pasta with Spinach-and-Walnut Pesto

RECIPE BY DR. ALKA CHANDNA

Conventional pesto is made with basil leaves, pine nuts, and cheese. "Taking out the cheese was a no-brainer," says Alka, "but as a graduate student in Canada, fresh basil wasn't always available and pine nuts were (and are) expensive. Using spinach and omega-3-rich walnuts as stand-ins gave birth to this recipe." In addition to tossing this with pasta, as in this recipe, you can also spread the pesto onto French bread, or use it as a dip with crackers or veggies.

SERVES 4 TO 6

3 cups (90 g) chopped fresh spinach

¼ cup (25 g) raw or toasted walnuts

3 cloves garlic, peeled

3 tablespoons olive oil

½ teaspoon salt

1 tablespoon fresh lemon juice (optional)

1 pound (455 g) any pasta

In a food processor, combine the spinach, walnuts, garlic, oil, salt, and lemon juice, if using, and process until smooth and creamy. Set aside.

Bring a large pot of salted water to a boil. Cook the pasta as the label directs. Drain, reserving 1 cup (240 ml) of the cooking liquid. Return the pasta to the pot, add the pesto, and toss to combine, adding some of the cooking liquid, a little at a time, to achieve the desired consistency. Serve immediately.

Mimi's Spaghetti

RECIPE BY KLAYTON RUTHERFORD

"I was thrilled to find out that with one simple switch, I could still eat Mimi's spaghetti, which has been my favorite dish for my entire life," says Klayton. "My mimi, like most mimis, is set in her ways as she approaches eighty, but she happily makes me a batch of vegan spaghetti when the menu calls for it, and everyone in the family has acknowledged that they can't tell the difference between the vegan and non-vegan versions."

SERVES 4 TO 6

2 tablespoons olive oil

½ cup (75 g) sliced onion

1 pound (455 g) vegan beefy crumbles

2 cloves garlic, finely chopped

2 (14½-ounce / 415-g) cans diced tomatoes

2 (8-ounce / 225-g) cans tomato sauce

¼ cup (13 g) chopped fresh parsley

1½ teaspoons dried oregano

1 teaspoon salt

¼ teaspoon dried thyme

1 bay leaf

1 pound (455 g) spaghetti

In a large saucepan, heat the oil over medium heat. Add the onion and cook until golden, about 5 minutes. Add the beefy crumbles and garlic; sauté lightly. Add the tomatoes, tomato sauce, parsley, oregano, salt, thyme, bay leaf, and 1 cup (240 ml) water and stir to combine. Cover and simmer for 2 hours, stirring occasionally. Remove the cover and simmer for another 30 minutes. Remove the bay leaf.

Bring a large pot of salted water to a boil. Cook the spaghetti as the label directs. Drain. Serve the sauce over the spaghetti. Store leftover sauce in a tightly sealed container in the refrigerator for up to 5 days or in the freezer for up to 1 month.

Soba Noodle Buddha Bowl

RECIPE BY KIMBERLY PARSONS

"Buddha bowls have become a go-to dish for yogis," says Kimberly. "Typically, this meal-size bowl is filled with simple, pure food and enjoyed with deep gratitude. To create your own Buddha bowl, find a unique, large bowl, which you can see as a symbol of nourishment and gratitude, and perhaps a pair of chopsticks to complete the mood." In addition to a broth that is warming and very aromatic, this bowl is filled with vibrantly colorful ingredients that should visually excite your taste buds.

SERVES 4

FOR THE BOWLS:

5 cups (1.2 L) mushroom stock

1 star anise

2 whole cardamom pods

1 tablespoon chopped, peeled fresh ginger

3½ ounces (100 g) soba noodles

8 ounces (225 g) fresh chestnut and enoki mushrooms or other mushrooms

1 cup (95 g) finely shredded red cabbage

1 cup (50 g) carrot, peeled into thin strips

1 cup (70 g) finely shredded savoy cabbage

2½ ounces (70 g) sugar snap peas

1 cup (90 g) bean sprouts

TO SERVE:

Juice of 1 lime

Fresh cilantro leaves

Fresh jalapeño chile, thinly sliced (optional)

Black and white sesame seeds

Lime wedges

Tamari

In a large saucepan, combine the stock, star anise, cardamom, and ginger and bring to a boil over high heat. Reduce the heat to medium and add the noodles and mushrooms. Simmer for 5 minutes, or until the noodles are cooked.

Divide the broth and noodles into 4 serving bowls, removing the cardamom pods and star anise if you wish.

Evenly divide the red cabbage, carrot, savoy cabbage, peas, and sprouts among the bowls, arranging the vegetables in piles around the edges of each bowl. Garnish with the lime juice, cilantro, jalapeño, if using, sesame seeds, and lime wedges. Serve immediately with chopsticks and tamari.

Artichoke Penne

RECIPE BY VICTORIA MORAN

"I'm half Italian—the culinary half. Pasta, olive oil, artichokes, garlic, and wine—it comforts me just to read this recipe!" says Victoria.

SERVES 4

Salt

12 ounces (340 g) penne pasta

2 tablespoons olive, avocado, or macadamia oil

1 medium onion, chopped

2 (9-ounce / 255-g) boxes frozen artichoke hearts, thawed and sliced

3 cloves garlic, chopped

½ cup (120 ml) white wine

2 tablespoons fresh lemon juice

1 tablespoon vegan butter or olive oil

½ cup (55 g) sundried tomatoes, reconstituted or oil-packed, thinly sliced

Fresh basil leaves, for garnish

Bring a large pot of salted water to a boil. Cook the pasta as the label directs. Drain, reserving ½ cup (120 ml) of the cooking liquid.

In a large skillet, heat the oil over medium heat. Add the onion and sauté for about 5 minutes, or until it begins to soften. Stir in the artichoke hearts and garlic. Stirring, add the wine. Reduce the heat to medium-low, cover, and cook for 5 minutes. Carefully stir in the pasta cooking liquid and lemon juice, then add the butter, tomatoes, and ½ teaspoon salt and stir to combine. Toss the pasta with the artichoke sauce. To serve, divide the pasta into 4 bowls and garnish with basil. Serve hot.

Peanut Tempeh Soba Bowls

RECIPE BY TOFURKY

Soba noodles are made with buckwheat, which is confusingly more closely related to rhubarb than wheat. But you don't need to know botany to enjoy this incredibly tasty, protein-packed concoction.

SERVES 4

FOR THE TEMPEH AND NOODLES:

¼ cup (60 ml) creamy peanut butter

2 tablespoons fresh lime juice

2 tablespoons soy sauce

2 teaspoons toasted sesame oil

2 cloves garlic, finely chopped

½ teaspoon red pepper flakes

1 (7-ounce/200-g) package Tofurky Sesame Garlic Tempeh or other tempeh

12 ounces (340 g) soba noodles

2 cups (180 g) broccoli florets

FOR THE SOY-SESAME SAUCE:

½ cup (120 ml) rice vinegar

¼ cup (60 ml) soy sauce

1 tablespoon fresh lemon juice

2 cloves garlic, minced

3 green onions, trimmed and thinly sliced

1 teaspoon toasted sesame oil

TO SERVE:

1 large avocado, pitted, peeled, and thinly sliced

2 large carrots, shredded or thinly sliced

Preheat the oven to 375° (190°C).

Make the tempeh: In a medium bowl, combine the peanut butter, lime juice, 2 tablespoons soy sauce, 2 teaspoons sesame oil, 2 cloves garlic, and the red pepper flakes and whisk with a wire whisk until smooth. Add the tempeh and gently toss to coat. On a baking sheet, spread out the tempeh in a single layer and bake until golden brown, 15 to 20 minutes.

Cook the noodles: Bring a large pot of water to a boil. Cook the noodles as the label directs. For the last 2 minutes of cooking, add the broccoli. Drain the broccoli and noodles and rinse under cold water until completely chilled.

Make the soy-sesame sauce: In a small bowl, whisk together the vinegar, ¼ cup (60 ml) soy sauce, the lemon juice, 2 cloves garlic, the green onions, and 1 teaspoon sesame oil.

To serve, divide the noodles and broccoli among 4 bowls and top with the tempeh, avocado, and carrots. Drizzle with the soy-sesame sauce.

Vegan Seafood Pasta

RECIPE BY EMILY LAVIERI-SCULL

"This dish is filled with lots of mushrooms and delicious umami flavors, similar to some of the seafood pasta dishes that I loved growing up," says Emily. "Replicating the flavors and textures of tuna, anchovies, and clams turned out to be relatively easy, and now I can enjoy my take on a childhood favorite." For this recipe, try a chunky pasta, such as cavatappi. Linguine works well, too.

SERVES 4

Salt

1 pound (455 g) any pasta

2 tablespoons vegan butter

2 tablespoons olive oil

1 medium onion, chopped

3 to 4 green onions, trimmed and thinly sliced

3 to 4 cloves garlic, finely chopped

1 pound (455 g) mixed mushrooms, such as cremini, shiitake, and oyster, roughly chopped

½ cup (25 g) thinly sliced sundried tomatoes

3 tablespoons capers

1 sheet nori, crumbled or finely chopped

Juice of 1 lemon

1 tablespoon liquid aminos or soy sauce

Ground black pepper

¼ to ½ teaspoon red pepper flakes (optional)

1 cup (30 g) chopped parsley

Nutritional yeast or vegan parmesan, for serving

Bring a large pot of salted water to a boil. Cook the pasta as the label directs. Drain, reserving ½ cup (120 ml) of the cooking liquid.

In a cast-iron pan or nonstick skillet, heat the butter and oil over medium heat. Add the onion and sauté for 5 minutes, or until it begins to soften. Add the green onions and stir, cooking until the onions start to brown, about 5 minutes.

Add the garlic and sauté for 1 minute. Add the mushrooms and stir occasionally, letting them sit so they brown and caramelize, about 5 minutes.

Add the tomatoes, capers, nori, lemon juice, liquid aminos, salt, pepper, and red pepper flakes, if using, and stir, letting the flavors come together, about 5 minutes. Start slowly adding some of the pasta cooking liquid to reach the desired consistency. Add the parsley right before folding the sauce in with the pasta. Top with nutritional yeast or vegan parmesan and serve warm.

If we keep fishing at the current rate, all species of wild seafood will **COLLAPSE** within 50 years.

Chapter Six

MAIN
DISHES

Bell Pepper Quiche

RECIPE BY PLANTPURE/KIM CAMPBELL

This oil-free quiche uses sweet bell peppers for the crust, which makes the preparation so much simpler. The filling is a creamy tofu-based "cheese" loaded with hearty potatoes, spinach, and asparagus. You can make them ahead and bake them when you're ready to eat. Drizzling vegan hollandaise sauce over the top adds some extra punch.

SERVES 3

1 (14-ounce / 400-g) package extra-firm tofu, drained

2 tablespoons tahini

2 teaspoons onion powder

1 teaspoon garlic powder

2 tablespoons cornstarch

⅓ cup (20 g) nutritional yeast flakes

½ teaspoon sea salt

¼ teaspoon ground black pepper

1 tablespoon chopped fresh thyme

1 tablespoon chopped fresh rosemary

2 cups (340 g) fresh or frozen shredded potatoes

2 cups (60 g) chopped fresh spinach

10 asparagus spears, trimmed and cut into ½-inch (12-mm) pieces

3 red, orange, or yellow bell peppers, seeded and halved

½ to ¾ cup (120 to 180 ml) Hollandaise Sauce (page 41)

Preheat the oven to 375°F (190°C). Line a 9 by 13-inch (23 by 33-cm) baking pan with parchment paper.

In a food processor, combine the tofu, tahini, onion powder, garlic powder, cornstarch, yeast flakes, salt, pepper, thyme, and rosemary. Process until the mixture is creamy, like the consistency of ricotta cheese.

Transfer the tofu mixture to a large bowl. Fold in the potatoes, spinach, and asparagus, mixing until well incorporated.

Place the bell pepper halves on the lined baking pan, cut side up. Generously fill each bell pepper half with the tofu filling. Bake for 30 minutes, or until the tops are golden brown. Remove from the oven and allow to set for 15 minutes. Drizzle each quiche with 1 to 2 tablespoons hollandaise sauce and serve, allowing two pepper halves per serving.

Creamy Grits, Beans, and Collards

RECIPE BY MARIA KOUTSOGIANNIS

"I suggest doubling this recipe, as it's so good and keeps well, plus it's easy and simple to clean up!" says Maria. "This combo is great for any time of day—I especially love eating this type of meal after the gym. You can also add it to pastas, wraps, or salads."

SERVES 4 TO 6

FOR THE GRITS:

1 cup (165 g) coarsely ground polenta, such as Bob's Red Mill

1 teaspoon concentrated vegetable base, such as Better Than Bouillon Seasoned Vegetable Base

Salt and ground black pepper

1 cup (125 g) grated vegan cheese, such as Parmesan

FOR THE BEANS:

1 (15-ounce / 430-g) can butter beans, rinsed and drained

Juice of 1 lime

Salt and ground black pepper

1 tablespoon hot sauce

1 heaping tablespoon coconut sugar

1 tablespoon soy sauce

1 tablespoon extra-virgin olive oil

1 teaspoon ground cumin

1 teaspoon smoked paprika

Pinch of ground turmeric

Pinch of cayenne pepper

FOR THE CORN:

3 cups (410 g) frozen corn kernels

1 cup (240 ml) coconut cream

Salt and ground black pepper

¼ cup (10 g) chopped fresh cilantro, plus extra cilantro leaves for garnish (optional)

Splash of lime juice

FOR THE COLLARDS:

1 bunch collards, spines and stems removed, coarsely chopped

Juice of 1 lemon

Splash of olive oil

Salt and ground black pepper

Make the grits: In a pot, combine the polenta, 3 cups (720 ml) water, the vegetable base, and salt and pepper. Bring to a boil over high heat, then reduce the heat to medium and simmer, stirring frequently, for 7 minutes, or until the polenta is cooked. Add a little more water if you prefer a looser consistency. Add the cheese and stir until melted. Taste and season with more salt and pepper, if needed.

Make the beans: In a medium saucepan, combine the beans, lime juice, salt, pepper, hot sauce, sugar, soy sauce, 1 tablespoon oil, the cumin, paprika, turmeric, cayenne, and 3 tablespoons water over low heat and cook, stirring frequently, for 10 to 12 minutes, or until the beans are soft.

Make the corn: In a cast-iron skillet over medium heat, cook the corn for 3 minutes, or until softened. You can cover the skillet and add some water to help the corn steam. Add the coconut cream, salt, pepper, cilantro, and lime juice. Bring the mixture to a boil, then reduce the heat to low and simmer, stirring frequently, for about 5 minutes.

Make the collards: In a large skillet, combine the collards, lemon juice, a splash of oil, and salt and pepper over medium heat; cover and cook for about 7 minutes, or until the collards are soft enough to chew but still a bit crunchy.

To serve, divide the grits into 4 to 6 shallow bowls, then spoon on some of the beans, corn, and collards. Store leftovers in separate tightly sealed containers in the refrigerator, where they will keep well for up to 3 days.

Vegetable Fried Rice

RECIPE BY JUST

Cold rice works best for fried rice, so plan ahead and cook your rice in advance, or keep it simple and use frozen rice.

SERVES 4

2 tablespoons vegetable oil (or any oil in your pantry used for cooking veggies), divided

2 shallots, finely chopped

2 cloves garlic, finely chopped

½ cup (75 g) fresh or frozen green peas

½ cup (70 g) fresh or frozen chopped carrot

½ cup (120 ml) Just Egg or other vegan egg alternative

4 cups (515 g) cooked, cold white jasmine rice

2 tablespoons soy sauce or tamari, plus more if needed

1 teaspoon sesame oil

2 green onions, trimmed and thinly sliced

In a wok or large skillet, heat 1 tablespoon vegetable oil over medium-high heat. Add the shallots and stir-fry for 2 minutes, or until softened, then add the garlic, peas, and carrot and stir-fry for 3 to 5 minutes, or until the vegetables are tender. Transfer to a bowl and set aside.

Into the same wok or skillet, add the egg and scramble over medium-low heat until just cooked through, about 4 minutes, breaking it up with a stiff spatula into fluffy, bite-size pieces. Add the remaining 1 tablespoon vegetable oil and the rice and stir until the rice is coated with oil.

Add the reserved cooked veggies, the soy sauce, and sesame oil and stir. Stir-fry until the rice is heated through. Taste and add more soy sauce, if needed, and garnish with the green onions. Serve hot. Store leftovers in a tightly sealed container in the refrigerator for up to 3 days.

GREEN PEAS

Runoff from nitrogen fertilizer is one of the hazards of agriculture—it pollutes water, causing overgrowth of algae that deplete oxygen, resulting in "dead zones." But green peas, like other plants in the legume family, have a talent for creating their own natural fertilizer. The pea is a "nitrogen-fixing" crop, meaning it cooperates with soil bacteria called rhizobia, which live in their roots, to pull nitrogen gas from the air and transform it into soil-enriching fertilizer that helps those pea vines grow and climb up your garden trellis. Peas also like to share: When their season ends, they release the nitrogen so other nearby plants can use it.[37]

Lentil Shepherd's Pie

RECIPE BY SAM TURNBULL

"This shepherd's pie is perfect for the holidays or a hearty weekend meal," says Sam. "The garlic mashed potatoes take this from ordinary to extraordinary!" This dish can be made ahead of time and reheated.

SERVES 4 TO 6

1 tablespoon olive oil

1 medium onion, chopped

2 carrots, chopped

1 pound (455 g) white mushrooms, sliced

4 cloves garlic, finely chopped

2 teaspoons dried thyme

1½ cups (290 g) brown or green lentils, or 2½ (15-ounce/430-g) cans cooked lentils (see Note)

3 cups (720 ml) vegetable broth, or 1 vegetable bouillon cube (see Note)

1 cup (145 g) fresh or frozen green peas

¼ cup (60 ml) Ravens' Barbecue Sauce (page 98) or other vegan barbecue sauce

2 tablespoons soy sauce

1 recipe Garlic Mashed Potatoes (recipe follows)

Preheat the oven to 425°F (220°C).

In a large skillet, heat the oil over medium-high heat. Add the onion, carrots, mushrooms, garlic, and thyme and sauté for 6 to 8 minutes, or until the vegetables are softened and beginning to brown. Transfer the vegetables to a bowl and set aside.

In the same skillet, combine the lentils and vegetable broth and bring to a boil over high heat. Reduce the heat to medium, cover, and simmer for 25 to 30 minutes, or until the broth is absorbed and the lentils are tender.

Add the vegetables back to the pan with the lentils, then add the peas, barbecue sauce, and soy sauce. Heat through. Scoop the mixture into a casserole dish or large ovenproof skillet, then spread the mashed potatoes on top.

Bake for 10 to 15 minutes, or until the pie is hot and bubbling around the edges.

Note: For a quicker prep, you can use canned cooked lentils. Just rinse and drain, crumble 1 vegetable bouillon cube in with the lentils, and stir to combine.

Garlic Mashed Potatoes

**MAKES ABOUT
8 CUPS (2.6 KG)**

**3 pounds (1.4 kg)
Yukon gold potatoes
(8 medium), chopped**

6 cloves garlic, crushed

**1 cup (240 ml) unsweet-
ened plant milk, plus
more if needed**

**1 cup (240 ml) full-fat
coconut milk, plus more
if needed**

2 teaspoons salt

In a large pot, cover the potatoes with water and bring to a boil over high heat. Boil for 15 minutes, or until they are fork-tender. Drain and return the potatoes to the pot.

In a medium saucepan, combine the garlic, plant milk, and coconut milk. Bring to a simmer over medium heat and cook for 10 minutes.

Mash the potatoes, add about half of the milk mixture, and stir to combine. Slowly add the remaining milk mixture until the desired consistency is reached. If you use all the milk mixture and you would still like the potatoes softer, just add more coconut milk or plant milk. Add the salt and serve.

Mushroom Stroganoff

RECIPE BY VICTORIA MORAN

"This recipe has never failed me at dinner parties," says Victoria. "A friend's husband once remarked while leaving: 'I'd already mapped out the McDonald's in the neighborhood, thinking I'd need something to eat on the way home, but this not only tasted amazing—it filled me up in a way I never thought could happen without meat.'" To make this recipe oil-free, sauté the vegetables in ¼ cup (60 ml) vegan dry red wine instead of using oil.

SERVES 6 TO 8

1 pound (455 g) soft or medium tofu, drained and coarsely chopped

¼ cup (60 ml) olive or avocado oil

1 medium yellow onion, finely diced

2 pounds (910 g) white mushrooms, quartered

3 cloves garlic, finely chopped

½ cup (65 g) whole-wheat flour or all-purpose gluten-free flour

½ teaspoon dry mustard

Salt and ground black pepper

4 cups (960 ml) hot vegetable broth

Cooked brown rice or thick noodles, for serving

In a blender, puree the tofu with enough water to make a creamy consistency, about 2½ to 3 cups (600 to 720 ml). Set aside.

In a large skillet, heat the oil over medium heat. Add the onion and mushrooms and cook, stirring, for 5 minutes, or until the vegetables begin to soften. Stir in the garlic, flour, dry mustard, and salt and pepper. Cook for 5 minutes, stirring occasionally.

Gradually stir in the hot broth and the tofu cream. Taste and season with more salt and pepper, if needed. Serve hot over brown rice or noodles.

Sweet Potatoes in African Peanut Curry

RECIPE BY DARSHANA THACKER

"This stew is so easy to make that it surprises everyone with its deliciousness," says Darshana. "It tastes great made with any vegetables, so change it to your preference."

SERVES 6

1 medium onion, finely chopped (about 2 cups/250 g)

2 red bell peppers, seeded and finely chopped (about 2 cups/290 g)

1 tablespoon minced garlic

1 tablespoon grated, peeled fresh ginger

2 pounds (910 g) sweet potatoes (2 medium), coarsely chopped

1 tablespoon ground cumin

1 tablespoon paprika

⅛ teaspoon red pepper flakes

2 tablespoons creamy peanut butter

Sea salt

3 cups (about 21 g) cooked whole grains, such as rice, quinoa, or sorghum

1 tablespoon crushed roasted peanuts (optional)

In a saucepan, combine the onion, bell peppers, garlic, ginger, and ½ cup (120 ml) water over medium heat and cook, stirring occasionally, for 10 minutes, or until the vegetables are tender. Add 1 to 2 tablespoons water if necessary to prevent the vegetables from sticking to the pan.

Add the sweet potatoes, cumin, paprika, red pepper flakes, and 2 cups (480 ml) water. Bring to a boil and cook, covered, for 15 to 20 minutes, or until the potatoes are tender.

In a small bowl, use a wire whisk to whisk the peanut butter and 2 tablespoons water into a smooth paste. Add this mixture to the pan and stir to combine. Season with salt and bring the curry to a quick boil.

Divide the grains among 6 individual serving bowls. Pour the curry over the grains and garnish with peanuts, if desired. Serve warm.

Savory Lentil Sausage

RECIPE BY AMY WEBSTER

"I served the sausage with penne pasta and pesto sauce to my non-vegan family," says Amy. "Everyone filled up on the dish and went back for seconds. Even better, my aunt begged me for the lentil sausage recipe." The wheat gluten and lentils make this dish especially high in protein.

MAKES 4 LARGE SAUSAGES

1 cup (200 g) cooked lentils, mashed

1 cup (240 ml) vegetable broth

1 tablespoon olive oil

2 tablespoons soy sauce

⅓ cup (20 g) nutritional yeast

1½ teaspoons garlic powder, or 2 cloves garlic, finely grated

1½ teaspoons fennel seeds, crushed or ground

½ teaspoon red pepper flakes

1½ teaspoons smoked paprika

1 teaspoon dried oregano

½ teaspoon dried thyme

½ teaspoon ground black pepper

⅛ teaspoon ground allspice

1¼ cups (150 g) vital wheat gluten

Pour 1 inch (2.5 cm) water into a pot fitted with a steamer basket. Bring to a boil over high heat. Arrange 4 (18-inch) square sheets of aluminum foil on a work surface.

In a large bowl, combine the lentils, broth, oil, soy sauce, yeast, garlic powder, fennel seeds, red pepper flakes, paprika, oregano, thyme, pepper, and allspice and stir. Add the wheat gluten and combine well (you may need to use your hands).

Divide the dough into 4 even parts. Place one part of the dough onto one sheet of foil and mold into about a 4-inch- (10-cm-) long log. Wrap the foil around the dough, then repeat with the remaining foil and dough, making 4 sausages. Place the sausages in the steamer basket and steam for 40 minutes (replenishing the boiling water if necessary), or until the sausages are firm. Enjoy immediately or store in a tightly sealed container in the refrigerator for up to 3 days or in the freezer for up to 1 month.

LENTILS

The versatile little lentil is an excellent global source of protein and fiber that's greener than meat. Producing lentils has only 1⁄48th of the climate impact of producing beef. And while beef is a water hog, slurping up nearly 30 gallons per gram of protein from farm to plate, lentils sip only about 5 gallons.

Small wonder, then, that the United Nations named 2016 the International Year of Pulses (pulses being part of the legume family). Climate change strains agriculture and global food security. Because pulses like lentils both adapt well to a changing climate and help ease its effects, the U.N. Food and Agriculture Organization (FAO) says, "Introducing pulses into crop production can be key to increasing resilience."

Like peas, lentils are nitrogen fixers, keeping the atmospheric nitrogen they need to grow in the soil. That means they need less or no chemical fertilizers, which can run off into waterways. According to the FAO, lentils have fixed 3 to 6 million tons of nitrogen in the soil and even help increase the diversity of microbe life in soil.[38]

Country-Fried Tofu

RECIPE BY AMY WEBSTER

"I began working with this recipe because I love to show people how good tofu can be," says Amy. "Since the main ingredient here is tofu, it's not only better for everyone (especially the animals), but healthier for you, too. You can also bake the tofu for 30 minutes in a 425°F (220°C) oven for an even healthier favorite." Serve it with braised greens and vegan mac and cheese, or roasted asparagus and potatoes.

SERVES 4

Vegetable oil, for the baking sheets and for frying

1 cup (125 g) all-purpose gluten-free flour or whole-wheat flour

1 cup (60 g) nutritional yeast

1½ teaspoons salt

1 teaspoon garlic powder

1 teaspoon onion powder

1 teaspoon ground black pepper

¼ cup (35 g) ground flaxseeds

¼ cup (60 ml) warm water

2 tablespoons yellow mustard

1½ cups (360 ml) vegetable broth, or 3 vegetable bouillon cubes dissolved in 1½ cups (360 ml) boiling water

2 (12-ounce / 340-g) packages firm tofu, frozen and thawed twice (see Note), drained, and pressed (see page 20)

1 cup (60 g) crushed cornflakes or panko breadcrumbs

Line two baking sheets with parchment paper, then drizzle with oil and spread the oil around. Set aside.

In a large bowl, combine the flour, yeast, salt, garlic powder, onion powder, and pepper and stir. Set aside.

In a separate bowl, combine the flaxseeds with the warm water and let set for 5 minutes. Add the mustard and 1 cup (240 ml) additional water to the flaxseed mixture.

Add ¼ cup (30 g) of the flour mixture to the flaxseed mixture and whisk with a wire whisk until smooth.

Pour the broth into a shallow bowl. Break the tofu into 14 to 18 pieces, letting the tofu tear naturally. Piece by piece, squeeze any remaining water out of the tofu, then place the tofu into the broth and let the tofu absorb the flavor and moisture for at least 15 minutes. Squeeze out the excess broth and place the tofu on a plate or cooling rack.

Roll each piece of tofu gently in the flour mixture and return to the plate or rack. Add the cornflakes to the flour mixture and stir well.

Dip each piece of tofu into the flaxseed mixture, then coat well with the cornflake mixture. Transfer the tofu to the baking sheets, making sure the pieces are not touching.

In a wok or heavy pan, heat 2 inches (5 cm) oil to 350° (175°C) and fry the tofu in small batches, turning occasionally for even frying, for about 5 minutes per piece, or until the tofu is golden brown. Be sure the oil temperature does not exceed 360°F (180°C). Transfer the tofu to a paper towel–lined plate or a rack. Serve hot.

Note: Freeze the tofu in the packages, then thaw, refreeze, and thaw again. This will make the tofu firmer and bring out the water.

Pumpkin Seed–Crusted Tofu Steaks

RECIPE BY CARYN HARTGLASS/GARY DE MATTEI

A crunchy flavorful crust made with pumpkin seeds elevates ordinary tofu into an extraordinary main dish. Top with Hollandaise Sauce (page 41) and serve with potatoes or rice and your favorite vegetables.

SERVES 4

¾ cup (100 g) raw pumpkin seeds

1 teaspoon onion powder

1 teaspoon garlic powder

1 teaspoon ground turmeric

¼ teaspoon ground black pepper

Pinch of salt

⅓ cup (75 ml) unsweetened plant milk

2 tablespoons tahini

1 tablespoon soy sauce, tamari, coconut aminos, or Bragg Liquid Aminos

15 to 16 ounces (430 to 455 g) firm tofu, drained, pressed (see page 20), and cut into 8 slices

Preheat the oven to 400°F (205°C). Line a baking sheet with parchment paper.

In a food processor, grind the pumpkin seeds into a meal. (Or grind the seeds in a spice grinder, in small batches.) Add the onion powder, garlic powder, turmeric, pepper, and salt and pulse to incorporate. Pour into a shallow bowl.

In a separate bowl, beat the milk into the tahini, a few tablespoons at a time, with a fork until smooth. Beat in the soy sauce.

Dip each slice of tofu in the tahini sauce and then in the pumpkin seed mixture. Make sure the tofu is well coated with the mixture. Place on the baking sheet in a single layer. Bake for 30 minutes, or until the tofu is nicely browned. Serve immediately or cool to room temperature, then cover and refrigerate for up to 3 days. To reheat, arrange the tofu on a baking sheet lined with parchment paper and bake in a preheated 400°F (205°C) oven until hot, about 15 minutes.

Almond-Rosemary-Crusted Tofu

RECIPE BY SAM TURNBULL

This crispy, crunchy, crusted tofu is the perfect vegan main dish. It is delicious served with a side of greens and a potato, or wonderful sliced and placed on top of a salad. Easy to make and even easier to enjoy! To make this recipe gluten-free, use gluten-free breadcrumbs and flour.

SERVES 4

½ cup (70 g) raw almonds

2 tablespoons chopped fresh rosemary (about 2 sprigs)

½ cup (62 g) dry breadcrumbs

1 teaspoon lemon zest

½ teaspoon salt

¼ teaspoon garlic powder

¼ teaspoon ground black pepper

½ cup (65 g) all-purpose flour

1 cup (240 ml) full-fat coconut milk or other plant milk

1 (12-ounce / 340-g) package extra-firm or firm tofu, drained, pressed (optional, see Note), and cut into 8 slices

Lemon slices, for serving

Preheat the oven to 400°F (205°C). Line a baking sheet with parchment paper.

In a food processor, combine the almonds and rosemary and pulse until the almonds are finely chopped but some larger chunks remain. Add the breadcrumbs, lemon zest, salt, garlic powder, and pepper and pulse to combine. (Alternatively, finely chop the almonds and rosemary with a knife and stir everything together in a bowl.)

Transfer the almond-rosemary mixture to a medium bowl. Place the flour in a separate medium bowl. Pour the coconut milk into a third medium bowl.

Dredge a slice of tofu in the flour. Shake off the excess, then dip the tofu into the coconut milk. Finally, coat the tofu in the almond-rosemary mixture. Place the tofu on the baking sheet and repeat with the remaining tofu. Bake for 18 to 23 minutes, or until the crust is lightly golden and crispy. Garnish with lemon slices and serve hot.

Note: You can press the tofu if you like (see page 20), but it isn't required. Pressing the tofu will make it a little firmer and chewier. Unpressed tofu is more tender.

Eggplant Rollatini with Spinach and Almond Ricotta Filling

RECIPE BY CARYN HARTGLASS/GARY DE MATTEI

Tender slices of fresh eggplant are battered with a gluten-free pumpkin seed "breading," baked (not fried), topped with a spinach and almond ricotta cheese filling, and hand rolled. "This recipe is wonderfully rich and satisfying and has become one of our very favorites to serve to guests," says Caryn.

SERVES 4 TO 6

1 large eggplant (2¼ to 2½ pounds / 1 to 1.2 kg), cut lengthwise into ¼-inch- (6-mm-) thick slices

Salt

1½ cups (200 g) raw pumpkin seeds

2 teaspoons onion powder

2 teaspoons garlic powder

2 teaspoons ground turmeric

½ teaspoon ground black pepper

⅓ cup (75 ml) unsweetened plant milk

2 tablespoons tahini

1 tablespoon soy sauce, tamari, coconut aminos, or Bragg Liquid Aminos

3 cups (720 ml) Classic Italian Marinara Sauce (recipe follows), or 1 (24-ounce / 720-ml) jar marinara sauce

1½ to 2 cups (360 to 480 ml) Spinach and Almond Ricotta Filling (recipe follows)

Vegan mozzarella or Cashew Cream (recipe follows)

Place a wire rack over a large baking sheet. Arrange the eggplant slices in a single layer on the rack. Sprinkle generously with salt. Set aside for 15 to 20 minutes, allowing the salt to draw out excess moisture and bitterness from the eggplant. Rinse off the salt from the eggplant and pat dry with a kitchen towel.

Preheat the oven to 400°F (205°C). Line a baking sheet with parchment paper.

In a food processor, grind the pumpkin seeds into a meal. (Or grind the seeds in a spice grinder, in small batches.) Add the onion powder, garlic powder, turmeric, and pepper and pulse to incorporate. Pour into a shallow bowl.

In a separate bowl, beat the milk into the tahini, a few tablespoons at a time, with a fork until smooth. Beat in the soy sauce.

Dip each slice of eggplant in the tahini sauce and then in the pumpkin seed mixture. Make sure the eggplant is well coated with the mixture. Place on the baking sheet in a single layer.

Bake for 10 minutes. Flip the eggplant slices over and bake for another 5 minutes, or until the eggplant is tender. Remove from the oven. Let sit for a few minutes until cool to the touch.

Spread about ¾ cup (180 ml) marinara on the bottom of a 9 by 13-inch (23 by 33-cm) baking pan.

Place an eggplant slice on a work surface and spoon 1 to 2 table-spoons (depending on the size of the slice) spinach and almond ricotta filling onto the wide end of the slice. Roll up the slice and place in the baking pan. Repeat with the remaining eggplant and filling.

Pour the remaining marinara over the eggplant rollatini. Sprinkle with vegan mozzarella or drizzle with cashew cream. Return to the hot oven and bake until hot and bubbly, 20 to 30 minutes. If not serving right away, bring to room temperature, then cover and refrigerate for up to 3 days. To reheat, bring to room temperature, then bake in a preheated 400°F (205°C) oven until hot, about 30 minutes.

Spinach and Almond Ricotta Filling

**MAKES ABOUT
3 CUPS (720 ML)**

**FOR THE ALMOND
RICOTTA:**

2 cups (285 g) raw almonds, soaked for at least 24 hours, drained, skins removed

1 tablespoon white miso

1 teaspoon salt

1 cup (240 ml) Cashew Cream (recipe follows)

FOR THE FILLING:

1 tablespoon olive oil

2 cloves garlic, finely chopped

2 shallots, finely chopped

1 cup (155 g) frozen chopped spinach, thawed and squeezed dry

½ cup (30 g) nutritional yeast

¼ cup (8 g) chopped fresh parsley

2 tablespoons Cashew Cream (recipe follows)

1 teaspoon ground nutmeg

1 teaspoon salt

½ teaspoon ground black pepper

Make the almond ricotta: In a high-powered blender with a tamper, combine the almonds, miso, 1 teaspoon salt, and 1 cup (240 ml) cashew cream and blend until smooth, using the tamper to push all the ingredients into the blade. Transfer the ricotta to a jar, seal, and refrigerate. Almond ricotta keeps well in the fridge for several days. It may also be frozen for up to 1 month.

Make the filling: In a skillet, heat the oil over medium heat. Add the garlic and shallots and cook, stirring occasionally, until tender, about 4 minutes. Set aside and cool to room temperature.

When the garlic and shallots have cooled, transfer to a food processor with the spinach, yeast, parsley, 2 tablespoons cashew cream, the nutmeg, 1 teaspoon salt, and the pepper and pulse until combined. Scrape down the sides of the bowl. Continue processing for about 1 minute, stopping periodically to scrape down the sides. Transfer to a glass container, cover, and refrigerate until ready to use. The filling will keep in the fridge for a few days. It also freezes well for up to 1 month.

Cashew Cream

**MAKES ABOUT
2 CUPS (480 ML)**

2 cups (240 g) raw
cashews

Place the cashews in a container with a lid, cover them with water, seal, and refrigerate for at least 24 hours. Drain and rinse the cashews, transfer to a blender, and blend with ½ cup (120 ml) water to give it the thickness of heavy cream.

Transfer to a jar or other container with a lid, cover, and refrigerate until ready to use. It will keep in the fridge for several days.

Classic Italian Marinara Sauce

**MAKES ABOUT
2 QUARTS (2 L)**

1 cup (240 ml) vegan dry
red wine, white wine, or
water

1 large onion, chopped

3 cloves garlic, finely
chopped

½ cup (25 g) chopped
fresh parsley

Leaves of 1 sprig rose-
mary, chopped

Salt and ground black
pepper

4 (15-ounce / 430-g) cans
whole tomatoes

2 (6-ounce / 170-g) cans
tomato paste

In a large saucepan, heat the wine over medium heat until it begins to simmer. Add the onion, garlic, parsley, rosemary, and salt and pepper to taste. Simmer, stirring occasionally, for about 15 minutes, or until the onion is tender and the wine is reduced by about three-quarters.

Stir in the tomatoes, tomato paste, and 2 cups (480 ml) water. Bring to a boil, then reduce the heat to low. Cover the pot and simmer the sauce for 2 hours, stirring occasionally. Taste and adjust the seasonings, if needed. If not using right away, let cool then transfer to a container with a lid, cover, and refrigerate for up to 1 week or store in the freezer for up to 1 month.

Enchilada Pie

RECIPE BY DR. ALKA CHANDNA

"I've long enjoyed the versatility afforded by layered dishes like lasagna and spinach pie and have spent hours standing in grocery store checkout lines imagining the marriage of different flavors and ingredients in layered casseroles," says Alka. "This easy recipe for enchilada pie is similarly versatile. The stripped-down version features black beans layered with tortillas and enchilada sauce—but you can build it up from there, adding soy chorizo or vegan beef crumbles, chopped green onions, peppers, jalapeños, zucchini, black olives, vegan cheese, and anything else you might imagine while waiting in a checkout line." Serve this dish with guacamole, fresh salsa, vegan sour cream, and tortilla chips to create a feast.

SERVES 6

1 tablespoon olive oil

1 medium yellow onion, chopped

2 bell peppers, seeded and diced

1 (12-ounce / 340-g) package soy chorizo

2 (15-ounce / 430-g) cans black beans, rinsed and drained

1 (15-ounce / 430-g) jar salsa

1 (12-ounce / 340-g) jar vegan enchilada sauce, such as Trader Joe's brand

10 to 12 (6- to 7-inch / 15- to 17- cm) corn or flour tortillas

1 cup (125 g) shredded vegan Pepper Jack cheese, such as Daiya Foods brand (optional)

Preheat the oven to 350°F (175°C).

In a large skillet, heat the oil over medium-high heat. Add the onion and bell peppers and cook, stirring occasionally, until the onion is translucent, about 10 minutes. Stir in the chorizo and cook for about 3 minutes, then add the beans. Use the back of a flat wooden spoon or a potato masher to mash the beans a little. Cook for 3 minutes more. Add the salsa. Reduce the heat to medium and cook, stirring occasionally, for 10 to 15 minutes. Remove from the heat.

In a baking dish, spread a thin layer of enchilada sauce. Place a layer of tortillas over the sauce. Layer some of the bean-chorizo mixture over the tortillas and sprinkle with some of the cheese, if using. Repeat the layering process—sauce, tortillas, bean-chorizo mixture, cheese—until you've used up all the bean-chorizo mixture. Finish with a layer of tortillas topped with enchilada sauce and cheese, if using. Bake for 30 minutes, or until bubbly. Serve hot.

Tuscan Seitan Meatballs with Anaheim Pepper Sauce

RECIPE BY SWEET EARTH

Serve these delicious meatballs and sauce over your favorite pasta or enjoy them on their own with some toasted bread.

SERVES 4

FOR THE SAUCE:

3 tablespoons olive oil

1 medium onion, chopped

2 large Anaheim chiles, seeded and chopped

2 medium tomatoes, chopped

2 tablespoons tomato paste

¼ teaspoon salt

⅛ teaspoon ground black pepper

1 cup (240 ml) vegetable stock

Make the sauce: In a large skillet, heat 3 tablespoons oil over medium heat. Add the chopped medium onion and cook until it is translucent, 5 to 7 minutes. Add the chiles and cook, stirring frequently, until they begin to soften but not brown, about 3 minutes. Reduce the heat to low, add the tomatoes, and cook until the liquid has reduced by one-third, about 10 minutes. Stir in the tomato paste and cook for 5 minutes. Season with the salt and pepper. Add the stock and stir well. Simmer for 5 minutes. Remove from the heat and set aside.

Make the meatballs: In a medium skillet, heat 2 tablespoons oil over medium heat. Add the mushrooms and cook, stirring, for 3 minutes.

Stir in the garlic and cook for 2 minutes. Add the chopped ½ small onion and cook, stirring, until translucent, about 5 minutes. Stir in the carrot and cook until softened, about 5 minutes. Transfer the vegetables to a bowl and allow to cool slightly.

Add the seitan, garbanzo beans, oats, ½ cup (120 ml) water, and the flaxseeds to the bowl and stir to combine. Add the ketchup, mustard, Worcestershire sauce, and parsley and stir well. The mixture should hold together. If it seems dry, add another teaspoon or two of water. Using a small (1½- to 2-inch/4 to 5 cm) ice cream scoop, form 10 to 12 uniform balls.

In a medium skillet, heat the remaining 3 tablespoons oil over medium heat. Cook the meatballs gently until they are golden brown on all sides, about 10 minutes.

Transfer the meatballs to the sauce. Bring the sauce to a simmer over low heat and cook the meatballs for 5 minutes, or until they are heated through. Do not stir too much to avoid breaking them up.

TOMATOES

What would a salad, pasta, or pizza be without tomatoes? We don't have to find out because tomatoes are eco-friendly. Their water use is one of the lowest of all plant foods (just 214 m^3 per ton of produce) and their climate footprint is relatively small: 1.3 kilograms of CO_2 per kilogram of tomato paste. Tomatoes can also be grown hydroponically, meaning without soil—hydroponic plants are raised in trays using a mineral-based growing medium instead. This saves plenty of soil, land, and water. But the "greenest" tomato is the one grown in your own backyard: Mass-production factors such as packaging and transportation create much of tomatoes' environmental "footprint," but you can easily raise a few vines in a planter for truly local produce.[39]

FOR THE MEATBALLS:

5 tablespoons (75 ml) olive oil, divided

2 ounces (55 g) shiitake mushrooms, finely chopped

2 cloves garlic, finely chopped

½ small onion, finely chopped

2 ounces (55 g) carrot, finely chopped

6 ounces (170 g) Sweet Earth Tuscan Savory Grounds or other seitan crumbles

½ cup (80 g) cooked garbanzo beans, finely chopped

½ cup (45 g) quick-cooking oats

2 tablespoons ground flaxseeds, soaked in 2 tablespoons water for 20 minutes

2 tablespoons ketchup

1 teaspoon grainy mustard

1 teaspoon vegan Worcestershire sauce

¼ cup (13 g) finely chopped fresh parsley

Lettuce Boats with Quinoa-and-Rice Burgers

RECIPE BY FRY'S FAMILY FOODS

Make the components for these lettuce boats ahead of time and plan to assemble them just before serving, as the lettuce will wilt and soften with the filling if left for too long.

SERVES 2

FOR THE PESTO:

1½ cups (45 g) basil leaves, roughly chopped

1½ cups (45 g) fresh cilantro, roughly chopped

⅓ cup (45 g) shelled pistachios

4 teaspoons nutritional yeast

4 teaspoons fresh lime juice

1½ cloves garlic, minced

1 green chile, seeded

¾ cup (180 ml) olive oil

Salt and ground black pepper

FOR THE VEGGIE SLAW:

¾ cup (70 g) shredded red cabbage

¾ cup (90 g) thinly sliced cucumber

2 green onions, trimmed and thinly sliced

2 teaspoons fresh lemon juice

2 teaspoons olive oil

Salt and ground black pepper

FOR THE BURGERS:

4 Fry's Quinoa and Rice Protein Burgers or other vegan burger patties

TO SERVE:

8 romaine lettuce leaves

Make the pesto: In a food processor, combine the basil, cilantro, pistachios, yeast, lime juice, garlic, and green chile and process until coarsely chopped. With the motor running, add ¾ cup (180 ml) oil in a thin stream. Process until smooth and season with salt and pepper.

Make the veggie slaw: In a bowl, combine the cabbage, cucumber, green onions, lemon juice, 2 teaspoons oil, salt, and pepper and toss well.

Cook the burgers: Heat a gas grill to high or heat charcoal briquettes in a charcoal grill until ash forms and they glow bright orange. Lightly brush the grill with vegetable oil, then grill the burgers until lightly charred, about 4 minutes per side, turning once. Remove from the grill and slice the burgers into strips. Alternatively, heat a skillet over medium-high heat and sear the burgers for 4 to 5 minutes per side.

To serve, dollop a little pesto into each romaine leaf. Add some veggie slaw and top with a few strips of burger.

Stuffed Tofu Turkey

RECIPE BY JULIA MURRAY

A perfect holiday feast, the pecan stuffing rolled up in a tofu "turkey" will shine on the center of the dinner table. Serve with Mushroom Gravy (page 146). Any stuffing you don't use to stuff the tofu can be baked in a casserole dish and served as a separate side dish.

SERVES 8

4 (12-ounce / 340-g) packages firm tofu, drained, pressed (see page 20), and coarsely chopped

2 vegetable bouillon cubes, divided

2 tablespoons nutritional yeast, divided

2 tablespoons onion powder, divided

2 tablespoons garlic powder, divided

2 tablespoons chopped fresh sage, divided

1 teaspoon salt, divided

2½ cups (500 g) Pecan Stuffing (recipe follows)

3 to 4 tablespoons Bragg Liquid Aminos, soy sauce, tamari, or coconut aminos

1 recipe Mushroom Gravy (recipe follows)

Preheat the oven to 400°F (205°C).

Lay out 2 sheets of aluminum foil, side by side, overlapping about 3 inches (7.5 cm), with a layer of parchment paper on top. The parchment should be about 15 by 13 inches (38 by 33 cm), with the foil as a border beneath it.

In a food processor, combine half the tofu, 1 bouillon cube, 1 tablespoon yeast, 1 tablespoon onion powder, 1 tablespoon garlic powder, 1 tablespoon sage, and ½ teaspoon salt and process for 1 minute, or until a dough-like ball starts to form. Transfer the mixture to the parchment paper. Process the remaining tofu, bouillon cube, yeast, onion powder, garlic powder, sage, and salt and add the mixture to the pile on the parchment.

Spread out the tofu mixture into a ½- to ¾-inch- (12-mm- to 2-cm-) thick rectangle (about 8 by 12 inches / 20 by 30.5 cm). Spoon the stuffing onto the middle of the tofu "turkey" in a lengthwise tube shape. Use the parchment to roll one side of the tofu turkey over the stuffing, then the other side, until the tofu turkey overlaps about 1 inch (2.5 cm). Peel the parchment away and fold in the ends of the tofu mixture, sealing the stuffing inside. Use your fingers to press the tofu mixture over any exposed stuffing and close any cracks.

Pour the liquid aminos over the top and use a brush to cover the whole tofu turkey. Wrap tightly with the parchment and foil, folding on top and twisting the ends. Bake in the oven for 1 hour.

Let cool for 15 minutes, then place in the refrigerator. When ready to serve, remove the tofu turkey from the refrigerator, unwrap the foil and parchment, cut the tofu turkey into 1-inch (2.5-cm) slices, wrap it back up, and bake for 20 minutes at 350°F (175°C). Serve with the mushroom gravy.

Pecan Stuffing

**MAKES 8 TO 10 CUPS
(1.6 TO 2 KG)**

8 cups (280 g) cubed
bread

1 cup (110 g) chopped
yellow onion

1½ cups (150 g) chopped
celery

7 to 8 white mushrooms,
chopped

1½ cups (360 ml) vege-
table broth, plus more if
needed

1 tablespoon dried thyme

1 tablespoon minced
fresh rosemary

2 tablespoons minced
fresh sage

½ teaspoon Himalayan
pink salt or sea salt

½ teaspoon ground black
pepper

½ cup (60 g) chopped
pecans

Preheat the oven to 350°F (175°C).

Spread the bread evenly on a baking sheet. Bake for about 20 minutes, stirring at the 10-minute mark, until the bread is dry.

In a large pot, oil-free sauté the onion, celery, and mushrooms (adding a bit of the broth if the vegetables start to stick) for 5 minutes, or
until they begin to soften. Add the thyme, rosemary, sage, salt, and
pepper. Stir, then add 1½ cups (360 ml) broth.

Add the bread and pecans and stir well. The stuffing should be
moist but not soaking wet. If your stuffing is still a little dry, keep
adding broth until it's moist (you shouldn't need more than 2 cups /
480 ml total).

Set aside 2½ cups (400 g) stuffing for the tofu turkey filling. Transfer
the remaining stuffing to a baking dish and bake for 25 minutes to
serve as a side dish.

Mushroom Gravy

**MAKES 3 CUPS
(720 ML)**

2 medium yellow onions, chopped

5 cups (350 g) sliced mushrooms, such as cremini, white, or wild

3 tablespoons white wine

3 tablespoons all-purpose gluten-free flour or whole-wheat flour

2 tablespoons tamari

1 teaspoon cider vinegar

1 teaspoon sea salt

Ground black pepper

Heat a large stainless-steel skillet over medium heat. (Stainless steel is better than nonstick for browning the onions.) Dry-sauté the onions, cooking them for 2 minutes, then adding a splash of water to pull the caramelization off the pan. Sauté until the onions are translucent and slightly browned, about 10 minutes, adding more splashes of water if they're sticking.

Add the mushrooms and cook until their water evaporates and they start to brown, about 5 minutes. Deglaze the pan with the wine, stirring. Add the flour. Stir and cook for 30 seconds. Slowly add 2 cups (480 ml) water, stirring constantly, then bring to a simmer. Add another ½ cup (120 ml) water. Cook for 3 to 4 minutes, or until the liquid reaches a gravy-like consistency, then add the tamari, vinegar, salt, and pepper. Serve hot.

Eating vegan for one day uses

1,500 fewer gallons of water,

which is enough to meet the daily indoor needs of approximately 15 people in the United States.

Chapter Seven

DESSERTS

Chocolate Chip Cookies

RECIPE BY FORA FOODS

This is the essential chocolate chip cookie recipe, with a sprinkle of sea salt and the richness of FabaButter. Chill the dough for easier scooping. Or forget baking it and enjoy this naturally egg-free cookie dough straight up.

MAKES ABOUT 2 DOZEN

3 tablespoons unsweetened almond milk

1 tablespoon ground golden flaxseeds

⅔ cup (135 g) beet sugar

¼ cup (55 g) dark brown sugar

⅔ cup (160 g) FabaButter or other vegan butter, at room temperature

2 teaspoons vanilla extract

1¾ cups (225 g) all-purpose flour

½ teaspoon baking soda

½ teaspoon salt

1 cup (182 g) vegan chocolate chips, divided

1 teaspoon flaky sea salt, such as Maldon salt

In a small bowl, whisk the milk and flaxseeds with a wire whisk until smooth. Chill in the refrigerator.

Meanwhile, in a large bowl, use a stand mixer or hand-held electric mixer to cream the beet sugar and brown sugar with the butter until fluffy and creamy. Beat in the chilled flaxseed mixture and vanilla.

In a separate bowl, sift the flour, baking soda, and ½ teaspoon salt. Add half the dry ingredients to the wet ingredients and use a rubber spatula to mix.

Fold in ¾ cup (135 g) chocolate chips and the remaining dry ingredients. Tightly cover the bowl and chill the dough for at least 1 hour or overnight.

Preheat the oven to 350°F (175°C). Line two baking sheets with parchment paper.

Scoop 2 tablespoons of dough and roll into a ball. Place on the baking sheet. Repeat with the remaining dough, spacing the balls 2 inches (5 cm) apart. Press a few chocolate chips on top of each ball and flatten with the palm of your hand. Sprinkle the flaky salt on top of each cookie.

Bake for 8 to 10 minutes, or until the cookies are slightly browned around the edges. Allow the cookies to cool on the baking sheets for 2 minutes. Store in a tightly sealed container for up to 1 week.

Betty's No-Bake Cookies

RECIPE BY VICTORIA MORAN

"Named for my mother-in-law, Betty, who believed if you can get perfect cookies without even turning on the oven, isn't that the best way to get cookies? And these are yummy," says Victoria. "While you can use any nondairy milk, for best results you should use something a little rich, thus the recommendation for soy. Any raw oats will work, but I've found that the cookies have the best texture with five-minute oats—not old-fashioned oats, but not quick-cooking oats, either."

**MAKES ABOUT
2 DOZEN**

2 cups (400 g) beet sugar

½ cup (50 g) cocoa powder

½ cup (120 ml) unsweet-ened soy milk or other plant milk

½ cup (120 ml) creamy peanut butter

1 teaspoon vanilla extract

2 cups (200 g) 5-minute oats

In a large saucepan, combine the sugar, cocoa powder, milk, peanut butter, and vanilla over medium heat and bring to a bubble, stirring.

Add the oats and stir well. Remove from the heat and spoon cookie-size dollops of the mixture onto waxed paper.

Let sit at room temperature for at least 4 hours or overnight to take on a cookie consistency. Store at room temperature in a tightly sealed container for up to 1 week.

Japanese Lemon Bars

RECIPE BY PLANTPURE/KIM CAMPBELL

Made with beautiful Japanese sweet potatoes, these frozen delights are creamy, sweet, and loaded with fresh lemon and coconut flavors. You can find Japanese sweet potatoes at Asian markets or gourmet grocers.

MAKES 16 (2-INCH/5-CM) BARS

FOR THE CRUST:

1 cup (90 g) old-fashioned oats

¾ cup (75 g) walnuts

½ cup (45 g) unsweetened coconut flakes

½ cup (65 g) pitted Medjool dates, soaked in warm water for 10 minutes and drained

1 teaspoon ground cinnamon

2 teaspoons lemon extract

2 tablespoons unsweetened applesauce

FOR THE FILLING:

2 large Japanese sweet potatoes (1½ to 2 pounds / 680 to 910 g), roughly diced

1 (13-ounce / 390-ml) can light coconut milk

½ cup (120 ml) maple syrup

Zest of 1 lemon

¼ cup (60 ml) fresh lemon juice (about 2 large lemons)

¼ cup (20 g) unsweetened coconut flakes, for garnish

Make the crust: Line an 8-inch (20-cm) square baking pan with parchment paper. In a food processor, pulse the oats, walnuts, ½ cup (45 g) coconut flakes, the dates, cinnamon, and lemon extract until the mixture has a finely ground, sticky consistency. Add the applesauce and continue pulsing until the mixture sticks together. Transfer to the baking pan and press into an even layer. Set aside.

Make the filling: Bring a pot of water to a boil over high heat. Add the sweet potatoes and cook until fork-tender, about 15 minutes. Drain and cool.

In a food processor, combine the sweet potatoes, coconut milk, maple syrup, lemon zest, and lemon juice. Process until creamy and smooth.

Spread the filling over the crust and smooth the top with a knife or spatula. Place in the freezer for 3 to 4 hours. Remove from the freezer and thaw for 15 to 20 minutes. Cut into 16 (2-inch/5-cm) squares, garnish with ¼ cup (20 g) coconut flakes, and serve. Store leftovers in a tightly sealed container in the freezer, where they will keep for up to 1 month.

Grasshopper Bars

RECIPE BY PLANTPURE/KIM CAMPBELL

These beautiful and decadent layered bars are a minty chocolate delight! They are loaded with the flavors of chocolate and mint, paired with the creamy texture of ripe avocado and banana. If you prefer a pie, simply build the recipe in a pie plate. For another variation, you can easily turn the green minty filling into a mint-chocolate filling by adding ¼ cup (25 g) cocoa powder.

MAKES 16 (2-INCH/5-CM) BARS

FOR THE CRUST:

1 cup (90 g) old-fashioned oats

½ cup (50 g) walnuts

½ cup (65 g) pitted Medjool dates, soaked in warm water for 10 minutes and drained

¼ cup (25 g) cocoa powder

1 teaspoon peppermint extract

Pinch of sea salt

FOR THE FILLING:

2 ripe medium avocados

1 medium banana

1 tablespoon fresh lemon juice

2 tablespoons crème de menthe

½ teaspoon peppermint extract

½ cup (120 ml) maple syrup

FOR THE TOPPING:

½ cup (90 g) vegan chocolate chips, melted

¼ cup (30 g) cacao nibs, or ¼ cup (20 g) unsweetened coconut flakes

Make the crust: Line an 8-inch (20-cm) square baking pan with parchment paper. In a food processor, combine the oats, walnuts, dates, cocoa powder, 1 teaspoon peppermint extract, and the salt and process until the mixture is mealy. Add 2 to 3 tablespoons water, 1 tablespoon at a time, until the mixture has a sticky consistency that will hold together for a crust. Be careful not to make it too wet, just enough to begin sticking together. Transfer the mixture to the baking pan and press into an even layer. Set aside.

Make the filling: In a high-powered blender, combine the avocados, banana, lemon juice, crème de menthe, ½ teaspoon peppermint extract, and the maple syrup and blend until the mixture is smooth and has the consistency of frosting. Spread the filling over the crust and smooth the top with a knife or spatula. Place in the freezer for 6 to 8 hours, or until completely frozen.

Make the topping: Remove the pan from the freezer. Pour the chocolate over the filling and spread evenly. Garnish with cacao nibs or coconut flakes. Return the pan to the freezer. When fully frozen and firm, slice into 16 (2-inch/5-cm) squares and serve cold. Store leftovers in a tightly sealed container in the freezer for up to 1 month.

Raw Vegan Tiramisu

RECIPE BY LENA KSANTI

Traditional tiramisu, the classic "pick-me-up" dessert of Italy, is loaded with dairy and sugar. This raw vegan version only *tastes* indulgent.

SERVES 4 TO 6

FOR THE CRUST:

1 cup (100 g) walnuts

1 cup (125 g) pitted dates

2 tablespoons cold-pressed coffee

FOR THE LADYFINGER LAYER:

1 cup (100 g) almond meal

1 cup (125 g) pitted dates

2 tablespoons cold-pressed coffee

1 teaspoon vanilla extract

1 tablespoon cacao powder, plus more for garnish

FOR THE CREAM:

2 cups (240 g) cashews, soaked at least 1 hour or overnight

2 tablespoons coconut oil

3 tablespoons maple syrup

1 tablespoon vanilla extract

Make the crust: Line an 8 by 4-inch (20 by 10-cm) pan with parchment paper. In a high-powered blender, blend the walnuts into a flour. Add 1 cup (125 g) dates and 2 tablespoons coffee and blend well. Transfer the mixture to the pan and press into an even layer.

Make the ladyfinger layer: Line a baking sheet with parchment paper. In a high-powered blender, combine the almond meal, 1 cup (125 g) dates, 2 tablespoons coffee, 1 teaspoon vanilla, and 1 tablespoon cacao powder and blend until smooth. Transfer to the baking sheet, form the mixture into an 8 by 4-inch (20 by 10-cm) rectangle, and place in the freezer.

Make the cream: In a high-powered blender, combine the cashews, oil, maple syrup, 1 tablespoon vanilla, and 1 cup (240 ml) water and blend until smooth and thick, adding more water if needed.

Spread half the cream onto the crust, smoothing the top with a knife or spatula. Remove the ladyfinger layer from the freezer and place on top of the cream. Spread the rest of the cream on top of the ladyfinger layer and refrigerate for 6 to 8 hours to harden.

Remove the tiramisu from the fridge and dust with cacao powder. Cut into squares and serve.

Golden Milk Rice Pudding

RECIPE BY DEVORAH BOWEN

Golden milk, traditionally a warm Indian beverage made with turmeric, is comforting and healthful! It translates deliciously into a lightly sweet dessert that's even great for breakfast.

SERVES 4

1¼ cups (300 ml) unsweetened cashew milk, divided

⅔ cup (125 g) Arborio rice

Pinch of salt

1 (13-ounce / 390-ml) can coconut milk

⅔ teaspoon ground cinnamon

⅓ teaspoon ground cardamom

1 teaspoon ground turmeric

Pinch of ground black pepper

2 tablespoons maple syrup, agave nectar, or vegan honey

1½ teaspoons vanilla extract

1 small piece fresh ginger

In a medium pot over medium heat, combine 1 cup (240 ml) of the cashew milk, the rice, and salt and heat to just under a boil. Reduce the heat to medium-low and simmer, uncovered, for about 10 minutes, stirring occasionally to keep the rice from sticking. When the cashew milk is almost absorbed, add the coconut milk and stir well to incorporate. Cover the pot and cook for 15 minutes, stirring occasionally.

Add the cinnamon, cardamom, turmeric, pepper, maple syrup, vanilla, and ginger, stir well, and cook for 10 minutes, or until most of the liquid is absorbed and the rice is creamy.

Remove the ginger and transfer the rice pudding to a glass or ceramic dish. Refrigerate for at least 45 minutes. Before serving, add the remaining ¼ cup (60 ml) cashew milk and stir well to make the rice pudding creamy again. Serve chilled or at room temperature.

Dark Chocolate Avocado Mousse with Sea Salt Coconut Whipped Cream

RECIPE BY NUTPODS

This mousse is made with nutrient-dense ingredients that promote your vitality and overall well-being. Think antioxidant-rich cacao (which acts as a neurotransmitter in the body and elevates your mood); satiating, brain-boosting fats like avocado and cashew butter; and dairy-free Nutpods to enhance the ultra-creamy texture. This is a decadent, rich dark chocolate mousse that's surprisingly healthy—it contains just seven clean ingredients and is free of gluten, dairy, and refined sugar.

SERVES 4

FOR THE MOUSSE:

1 large, very ripe avocado, pitted and peeled

¾ cup (180 ml) Nutpods dairy-free unsweetened coffee creamer (original flavor) or other vegan creamer, plus more if needed

½ cup (120 ml) creamy cashew butter

½ cup (60 g) cacao powder

¼ cup (30 g) pitted dates, chopped, soaked in warm water for 10 minutes, and drained

¼ cup (60 ml) coconut nectar or monk fruit sweetener

½ teaspoon sea salt

Cacao nibs, for garnish (optional)

FOR THE WHIPPED CREAM:

1 (13-ounce / 390-ml) can coconut cream or coconut milk (see Note), chilled at least 24 hours

2 tablespoons coconut nectar or monk fruit sweetener

½ teaspoon sea salt

Make the mousse: In a high-powered blender, combine the avocado, ¾ cup (180 ml) creamer, the cashew butter, cacao, dates, ¼ cup (60 ml) coconut nectar, and ½ teaspoon salt and blend until smooth and creamy, adding more creamer if needed to reach a smooth consistency. Spoon the mousse into 4 mason jars or ramekins. Place in the freezer for 2 hours to chill.

Make the whipped cream: In a blender, or using a stand mixer or hand-held electric mixer, whip the coconut cream with 2 tablespoons coconut nectar and ½ teaspoon salt until firm. Refrigerate for 2 hours.

When ready to serve, top the mousse with the whipped cream and garnish with cacao nibs, if using. If not using right away, store the mousse and whipped cream in separate tightly sealed containers in the refrigerator for up to 3 days.

Note: If you are using coconut milk, scoop out the solid coconut and discard the liquid before whipping.

Chocolate-Maca Quinoa Bars

RECIPE BY KIMBERLY PARSONS

Maca is known as an adaptogen. These special kinds of herbs adapt to a variety of conditions within the body and help restore it to a healthy balance. Maca in particular works on the endocrine system to balance hormones in both men and women, increasing sexual desire. Safe to use daily, maca can be added to many recipes such as this one, chewy tahini-and-almond cookies, or your morning smoothies. These bars are best eaten from the fridge. Other puffed grains, such as millet, rice, spelt, or buckwheat may replace the quinoa.

MAKES 12 BARS

½ cup (120 ml) coconut oil

½ cup (120 ml) coconut nectar, vegan honey, or maple syrup

½ cup (60 g) cacao powder

⅓ cup (40 g) maca powder

Pinch of salt, plus more if needed

3 cups (211 g) puffed quinoa

½ cup (70 g) roasted hazelnuts, roughly chopped

¼ cup (35 g) dried unsweetened cranberries, roughly chopped

¼ cup (30 g) raw shelled pistachios, roughly chopped

Line an 8-inch (20-cm) square baking dish with parchment paper.

In a medium saucepan, melt the oil over medium-low heat. Add the coconut nectar and stir well. Add the cacao, maca, and salt and whisk with a wire whisk until a loose paste forms. Remove from the heat.

Add the quinoa and stir to combine, making sure to coat all the quinoa with the chocolate paste. Taste and add more salt if needed.

Transfer the quinoa mixture to the baking dish. Use your palms to press it loosely into an even layer. Sprinkle the hazelnuts, cranberries, and pistachios evenly over the quinoa mixture and, using your palms, press the ingredients firmly into the baking dish. Refrigerate for 30 minutes, then cut into 12 bars. Store in a tightly sealed container in the fridge for up to 1 week, or in the freezer for up to 2 months.

HAZELNUTS

Hazelnuts grow on trees that climb to 15 feet tall, meaning they have the ability to pull and store carbon dioxide from the atmosphere—and they are particularly good at that task. Though most hazelnuts are grown in Turkey, you can save some "food miles" (the distance food travels during shipping and the related emissions and pollutants) by buying American: Oregon is a major global hazelnut producer.

This easy-to-maintain perennial goes relatively light on water use compared with other nuts, can be grown in almost any type of soil, and is drought-resistant due to its extensive roots, which also help build up the soil, retain nitrogen, and sequester more carbon than annual crops.

Strawberry Shortcakes

RECIPE BY FRAN COSTIGAN

Don't be put off by the long recipe: Each of the following components is quick and easy and can be made ahead of time and assembled before serving. "The luscious, silken vanilla cream turns a simple dessert into a fancy one and freezes beautifully, too," says Fran. "In addition to using the cream in this recipe, you can spoon it over fresh fruit, pudding, or gelled desserts."

MAKES 12 SHORTCAKES

1⅓ cups (315 ml) full-fat coconut milk

1 teaspoon vanilla extract

¼ teaspoon almond extract (optional)

Zest of 1 orange (about 1½ tablespoons)

¾ cup (95 g) all-purpose flour

½ cup (60 g) whole-wheat pastry flour

1 tablespoon baking powder

1¼ cups (125 g) almond meal

½ teaspoon fine sea salt

2 tablespoons beet sugar or coconut sugar (optional)

Lemon-Scented Silken Vanilla Cream (recipe follows)

Sweet-Tart Roasted Strawberries (recipe follows)

Sweet Balsamic Reduction (recipe follows)

Position a rack in the center of the oven and preheat to 450°F (230°C). Line a heavy baking sheet with a double layer of parchment paper.

In a measuring cup, combine the coconut milk, vanilla, almond extract, if using, and orange zest. Stir and set aside.

In a medium bowl, sift the all-purpose flour, pastry flour, baking powder, almond meal, and salt. Stir to combine.

Make a well in the center of the dry ingredients and pour in the coconut milk mixture. Stir, bringing the dry ingredients into the liquid, rotating the bowl as you go. Do not press or overmix. It comes together quickly into a craggy, soft dough.

Scoop a scant ¼ cup (60 ml) of dough per biscuit onto the baking sheet, leaving 2 inches (5 cm) in between. For ease, use a ¼-cup (60-ml) ice cream scoop. Sprinkle with the sugar, if using.

Bake for 14 to 15 minutes, or until set and very lightly golden. Flip one over; the bottom should be brown. Cool on a wire rack for a couple of minutes.

To serve, cut the biscuits in half using a serrated knife and spoon on some of the vanilla cream and strawberries.

The biscuits can be stored in a tightly sealed container in the freezer for up to a month. Defrost covered, and warm in a low oven.

Lemon-Scented Silken Vanilla Cream

MAKES 1⅓ CUPS (315 ML)

1 cup (120 g) raw cashews, soaked at least 1 hour or overnight

1 cup (240 ml) full-fat coconut milk

4 to 6 tablespoons (75 g) beet sugar

1 tablespoon neutral oil

⅛ teaspoon fine sea salt

Seeds from 1 vanilla bean, or 1 tablespoon vanilla extract

Zest of 1 lemon

In a high-powered blender starting on low, blend the cashews, coconut milk, 4 tablespoons (50 g) sugar, the oil, and salt. Increase the speed to high and blend for 2 minutes, or until the mixture is smooth, thick, and warm. It will be quite thick, and easiest to remove while warm. Taste for sweetness and add as much of the remaining 2 tablespoons of sugar as needed.

Pour the cream into a container. Stir in the vanilla and lemon zest.

Use the cream immediately or cover and refrigerate for 2 hours to overnight to allow the flavors to blend. The cream will thicken as it chills. Stir before using. The cream will keep well in the refrigerator for up to 3 days.

Sweet-Tart Roasted Strawberries

**MAKES ABOUT
1½ CUPS (485 G)**

1 quart (575 g) strawber-
ries, hulled

2 tablespoons maple
syrup

2 tablespoons Sweet
Balsamic Reduction
(recipe follows)

1 tablespoon framboise
(optional)

Position a rack in the center of the oven and preheat to 375°F (190°C).

Cut the smaller berries in half and the larger ones into quarters and
transfer to a medium bowl. In a small bowl, combine the maple syrup
and balsamic reduction and stir. Pour over the berries and toss to
coat.

Spoon the berries and the liquid into a 9 by 13-inch (23 by 33-cm)
baking pan, arranging the berries in a single layer.

Roast for 6 to 8 minutes, or until the berries are softened but not
mushy. Remove from the oven.

Set a strainer over a bowl. Drain the berries and spoon them into a
container.

In a small saucepan over high heat, pour the cooking liquid and
reduce until syrupy, about 5 minutes. Pour into a small bowl and
refrigerate until needed. Put the berries into a serving dish. Expect
more liquid to accumulate in the dish. When ready to serve, add the
syrup and framboise, if using, to the berries. The roasted berries are
best the day they are made, but they can be refrigerated overnight.

Sweet Balsamic Reduction

**MAKES ABOUT ½ CUP
(120 ML)**

1 cup (240 ml) balsamic
vinegar

3 tablespoons light agave
nectar

½ teaspoon lucuma
powder (optional)

In a small saucepan, combine the vinegar and agave. Bring to a
low boil over medium heat, whisking a few times with a wire whisk.
Reduce the heat to low and simmer, whisking occasionally, until
the syrup is reduced to ½ cup (120 ml), about 10 minutes. Stir in
the lucuma powder, if using. Properly made, the reduction will have
thickened enough to drip very slowly off a spoon.

Barely Any Sugar Blueberry Cobbler

RECIPE BY SHARON GANNON

This cobbler is best served while still warm. For an added treat, serve with a scoop of vegan vanilla ice cream.

SERVES 6

FOR THE FRUIT FILLING:

4 cups (580 g) fresh or frozen blueberries

1 cup (130 g) prunes, chopped

2 teaspoons vanilla extract

2 tablespoons all-purpose flour

1 tablespoon cornstarch

FOR THE BISCUIT TOPPING:

1½ cups (190 g) all-purpose flour

1 cup (85 g) quick-cooking oats

2 tablespoons beet sugar

1½ teaspoons baking powder

½ teaspoon salt

6 tablespoons (84 g) cold vegan butter

¾ cup (180 ml) plant milk

Preheat the oven to 375°F (190°C).

Make the fruit filling: In a medium bowl, combine the blueberries, prunes, and vanilla and stir well. Sprinkle 2 tablespoons flour and the cornstarch over the mixture and stir until the flour coats the fruit. Spread evenly into an ungreased 8-inch (20-cm) square baking dish or an 8 by 10-inch (20 by 25-cm) glass or porcelain baking dish, at least 2 inches (5 cm) deep. Do not use a metal baking pan. Set aside.

Make the biscuit topping: In a large bowl, combine 1½ cups (190 g) flour, the oats, sugar, baking powder, and salt. Stir until thoroughly mixed. Using two butter knives or a pastry cutter, cut in the butter. The mixture should look pebbly. Stir in the milk, making a sticky dough. Using your hands, form the dough into small biscuit-like patties, about ½-inch thick by 2 to 3 inches in diameter, and lightly press on top of the fruit filling until you cover all of it.

Bake for 45 minutes, or until the biscuit topping is golden brown. Cool for 15 minutes before serving to allow the filling to set.

HEMP

Hemp milk's growth in popularity as a cow's milk alternative and hemp seed oil's discovery by cooks are only the latest in a long line of eco-friendly uses for hemp. Not to be confused with THC-containing marijuana or CBD oil, although it comes from the same plant, hemp is organically grown and helps enrich poor soils. Its non-food parts are feedstock for a sturdy form of cloth (a substitute for pesticide-heavy cotton); paper (instead of chopping down trees); "hempcrete," a non-chemical product used to insulate buildings; non-petroleum plastics; nontoxic fuel (biodiesel); paint and varnish; and other products. America's first president, George Washington, swore by it, and it's once again being planted at his former home, Mount Vernon.

Hemp is also a hero in an environmental cleanup process called phytoremediation, in which certain plants serve to sop up water and soil pollution. It was even planted to help reduce heavy-metal contaminants around the site of the Chernobyl nuclear disaster.[40]

Strawberry Almond Cookies

RECIPE BY SARA KIDD

"These little cookies are bursting with color and flavors: strawberry, almond, and white chocolate—oh my!" says Sara. "They're a true crowd-pleaser and the authentic strawberry taste is such a surprise. Super-easy to bake and you will want to make them more than once!"

MAKES 2 DOZEN

FOR THE ALMONDS:

2 tablespoons vegetable oil

½ cup (55 g) slivered almonds

¼ cup (60 ml) maple syrup

½ tablespoon ground cinnamon

½ tablespoon vanilla extract

½ teaspoon sea salt

FOR THE COOKIE DOUGH:

1½ cups (195 g) cake flour, plus more if needed

½ teaspoon sea salt

¼ teaspoon baking powder

½ tablespoon ground cinnamon

½ cup (9 g) freeze-dried strawberries, crushed into small pieces

½ cup (100 g) beet sugar

¼ cup (55 g) light brown sugar

½ cup (112 g) vegan butter, at room temperature

Make the almonds: In a skillet over medium-high heat, warm the oil and add the almonds, maple syrup, ½ tablespoon cinnamon, ½ tablespoon vanilla, and ½ teaspoon sea salt. Cook, stirring, until the mixture starts to caramelize and the nuts are sticking together, about 4 minutes. It should be a dark caramel color. Remove from the heat and set aside.

Make the cookie dough: Preheat the oven to 350°F (175°C). Grease two baking sheets.

In a medium bowl, sift the flour, ½ teaspoon sea salt, the baking powder, and ½ tablespoon cinnamon, then stir in the strawberries.

In a large bowl, use a stand mixer or hand-held electric mixer to cream the beet sugar and brown sugar with the butter until fluffy and creamy. Add the cornstarch, 3 tablespoons vanilla, and the almond extract and mix on low speed until combined.

Add the dry ingredients to the wet ingredients and use your hands (see Note) to mix until just combined into a dough that sticks together. Don't overmix. If the dough is too wet, add more flour until the dough is dry but still holds its shape. Add the white chocolate chips and stir until just combined.

Using your hands, roll about 1 tablespoon of dough into a ball and place onto the baking sheet. Repeat with the remaining dough, spacing the balls 2 inches (5 cm) apart. Using a teaspoon, make a dent in the middle of each ball and fill with the almonds, pushing them into the dough. Bake for 10 to 14 minutes, or until the cookies are firm to the touch. Don't overbake. Remove from the oven and let cool.

Make the icing: In a small bowl, combine the melted chocolate and food coloring, if using, then drizzle the icing over the cooled cookies using a teaspoon. Garnish each cookie with a tiny amount of flaky salt and serve. Store leftover cookies in a tightly sealed container for up to 5 days.

Note: If you're using a stand mixer, you can mix on very low speed with the paddle attachment instead of using your hands.

2 tablespoons cornstarch mixed with 2 tablespoons cold water

3 tablespoons vanilla extract, or 2 tablespoons vanilla bean paste

1 teaspoon almond extract

½ cup (85 g) vegan white chocolate chip

FOR THE ICING:

¼ cup (45 g) vegan white chocolate chips, melted

Oil-based pink food coloring (optional)

Flaky sea salt, for garnish

Mango Chia Seed Pudding

RECIPE BY PEGGY CHAN

"This pudding is best when made with mangoes that are sweet and perfectly ripe," says Peggy. Serve this refreshing pudding with granola, puffed quinoa, or diced fresh fruits.

SERVES 4 TO 6

3 cups (495 g) chopped fresh mango

1 tablespoon maple syrup or coconut nectar

½ cup (120 ml) coconut milk

2 tablespoons chia seeds

In a high-powered blender, combine the mango, maple syrup, and coconut milk and puree until smooth. Transfer to a bowl and fold in the chia seeds. Cover and refrigerate overnight, and keep refrigerated until ready to serve. The pudding will keep well in the refrigerator for up to 3 days.

Chocolate Pie

RECIPE BY DR. ALKA CHANDNA

"The famed gourmet vegan restaurant Millennium used to have an insanely decadent dessert called Chocolate Midnight, which involved a cashew crust, chocolate filling, and various other ingredients," says Alka. "It took a day to make. When I started working at PETA, I no longer had a day to make a cake, so I stripped down the recipe and now I make this version."

SERVES 6 TO 8

2 (13-ounce / 370-g) packages silken tofu, such as Mori-Nu brand, drained

½ cup (100 g) turbinado sugar or beet sugar

1 teaspoon vanilla extract

½ teaspoon salt

1 cup (168 g) vegan chocolate chips, melted

1 Keebler Graham Pie Crust or other vegan graham-cracker pie crust

Preheat the oven to 350°F (175°C).

In a food processor, process the tofu until smooth, about 3 minutes, stopping periodically to scrape down the sides. Add the sugar, vanilla, and salt and process for 2 minutes. Add the chocolate and process for 2 minutes, or until silky smooth.

Pour the chocolate mousse into the crust and bake for 40 minutes, or until the mousse is set. Remove from the oven and let cool. Refrigerate for at least 3 hours or overnight (longer is better). Slice and serve cold. Store leftovers tightly covered in the refrigerator for up to 3 days.

Fudgy Chocolate Brownies

RECIPE BY RIPPLE FOODS

These brownies are chocolaty, rich, and delicious—everything you want in a fudgy brownie.

MAKES 18 BROWNIES

2 cups (255 g) all-purpose flour

2 cups (400 g) beet sugar

¾ cup (70 g) cocoa powder

1 teaspoon baking powder

1 teaspoon salt

1 cup (240 ml) Ripple Unsweetened Vanilla Pea Milk or other unsweetened vanilla plant milk

1 cup (240 ml) coconut oil, melted

1 tablespoon vanilla extract

1 cup (120 g) chopped walnuts (optional)

Preheat the oven to 350°F (175°C). Grease a 9 by 13-inch (23 by 33-cm) baking pan.

In a bowl, combine the flour, sugar, cocoa powder, baking powder, and salt.

In a small bowl, whisk the milk, oil, and vanilla. Pour into the flour mixture and stir until well blended. Fold in the walnuts, if using.

Transfer the batter to the baking pan. Bake for 28 to 32 minutes, or until firm. Cool for at least 20 minutes before slicing into 18 brownies. Store leftovers tightly covered in the refrigerator for up to 3 days or in the freezer for up to 1 month.

Vegan Brownie Parfaits

RECIPE BY RIPPLE FOODS

Here's a fun and decadent way to serve your next batch of brownies: Layer them in small jars or glasses with vegan chocolate pudding and coconut whipped cream.

SERVES 8

1 (1.7-ounce / 50-g) box vegan chocolate pudding mix, such as Jell-O brand

1⅔ cups (405 ml) Ripple Unsweetened Vanilla Pea Milk or other unsweetened vanilla plant milk

1 recipe Fudgy Chocolate Brownies (page 168)

1 (9-ounce) container coconut whipped cream

Prepare the chocolate pudding with the milk according to the package directions. Cut the brownies into bite-size squares. In small jars or glasses, layer the brownie bites, pudding, and whipped cream. Repeat the layers, ending with whipped cream. Serve immediately or cover and refrigerate until needed. These are best if served on the same day they are assembled.

White Chocolate Candy Bars

RECIPE BY JUSTINE DROSDOVECH

Coconut, caramel, and white chocolate combine for a decadent interpretation of the frozen Snickers bars we enjoyed as children. Be sure to let the can of coconut milk sit in the fridge overnight, then scoop the top layer off for the coconut milk in this recipe.

MAKES 16 BARS

FOR THE CRUST:

¾ cup (65 g) old-fashioned oats

⅔ cup (75 g) almond flour

⅓ cup (30 g) unsweetened coconut flakes

Himalayan pink salt

½ cup (65 g) pitted dates, soaked in warm water for 10 minutes and drained

FOR THE CARAMEL LAYER:

¾ cup (95 g) pitted dates, soaked in warm water for 10 minutes and drained, ¼ cup (60 ml) soaking water reserved

¼ cup (60 ml) creamy cashew butter

1 teaspoon vanilla extract

Pinch of ground cinnamon

½ cup (70 g) chopped salted peanuts

FOR THE WHITE CHOCOLATE LAYER:

½ cup (56 g) coconut manna, melted

3 tablespoons full-fat coconut milk

1 tablespoon coconut flour

2½ tablespoons maple syrup

½ teaspoon vanilla extract

Make the crust: Line an 8-inch (20-cm) square baking pan with parchment paper. In a food processor, combine the oats, flour, shredded coconut, and salt, then add ½ cup (65 g) dates and process to form a dough. Transfer the dough to the pan and press into an even layer. Place in the freezer.

Make the caramel layer: In a food processor or blender, combine ¾ cup (95 g) dates, the reserved soaking water, the cashew butter, 1 teaspoon vanilla, and the cinnamon and process until smooth. Remove the pan from the freezer and spread the caramel evenly on top of the crust. Layer the peanuts on top of the caramel and return the pan to the freezer for 3 hours.

Make the white chocolate layer: In a bowl, combine the coconut manna, coconut milk, flour, maple syrup, and ½ teaspoon vanilla and stir well. Remove the pan from the freezer and spread the coconut mixture evenly on top of the caramel layer. Return the pan to the freezer for 3 hours, or until set, then cut into 16 bars. Store in a tightly sealed container in the freezer for up to 1 week.

New York–Style Cheesecake

RECIPE BY MIYOKO SCHINNER

"This rich and dense New York–style cheesecake is so simple to make, yet so delicious," says Miyoko. Serve this culinary crowd-pleaser unadorned or top with your favorite fresh berries or sliced fruit.

SERVES 8

FOR THE CRUST:

½ cup (60 g) walnut pieces

¾ cup (65 g) old-fashioned oats

3 tablespoons beet sugar

⅓ cup (75 ml) melted Miyoko's European-Style Cultured Vegan Butter or other vegan butter

1 teaspoon vanilla extract

½ teaspoon ground cinnamon

¼ teaspoon ground nutmeg

FOR THE CHEESECAKE:

2 pounds (910 g) Miyoko's Plainly Classic Vegan Cream Cheese or other vegan cream cheese, at room temperature

1 cup (200 g) beet sugar

¾ cup (180 ml) coconut cream

¼ cup (30 g) cornstarch or arrowroot

¼ cup (60 ml) maple syrup

1 tablespoon agar powder

1 tablespoon vanilla extract

2 teaspoons lemon zest

Make the crust: Line an 8- or 9-inch (20- or 23-cm) springform pan with parchment paper. In a food processor, combine the walnuts, oats, 3 tablespoons sugar, the butter, 1 teaspoon vanilla, the cinnamon, and nutmeg and process until crumbly. Transfer the mixture to the pan and press into an even layer.

Make the cheesecake: Preheat the oven to 350°F (175°C).

In a food processor, combine the cream cheese, 1 cup (200 g) sugar, the coconut cream, cornstarch, maple syrup, agar, 1 tablespoon vanilla, and the lemon zest and process until smooth and creamy, about 1 minute. Alternatively, in a large bowl, combine the cream cheese, 1 cup (200 g) sugar, the coconut cream, cornstarch, and maple syrup and use a hand-held electric mixer or a wooden spoon to mix well. Add the agar, 1 tablespoon vanilla, and the lemon zest and mix well.

Pour the batter into the pan. Bake for about 50 minutes, or until lightly browned on top. Let cool completely, then cover and refrigerate for several hours or overnight before serving. The cheesecake will keep well in the refrigerator for up to 3 days.

Lemon Cheesecake Frozen Cookies

RECIPE BY CARYN HARTGLASS/GARY DE MATTEI

Made with almonds and Brazil nuts and filled with a lemon cheesecake filling, these frozen cookies are raw and gluten-free. "Make double the recipe—you'll be glad you did!" says Caryn.

MAKES ABOUT 18 COOKIES

FOR THE LEMON CHEESECAKE FILLING:

1½ cups (180 g) raw cashews, soaked at least 1 hour or overnight

¾ teaspoon lemon zest

6 tablespoons (90 ml) fresh lemon juice (about 3 large lemons)

¼ cup (60 ml) maple syrup

FOR THE COOKIE DOUGH:

⅔ cup (95 g) almonds

⅓ cup (45 g) Brazil nuts

1 cup (130 g) prunes

1 teaspoon vanilla extract

Make the lemon cheesecake filling: In a food processor or a high-powered blender, process the cashews, lemon zest, lemon juice, and maple syrup until smooth. Place in a bowl.

Make the cookie dough: In a food processor, pulse the almonds, Brazil nuts, prunes, and vanilla until the mixture clumps together in a ball.

Cut two pieces of parchment paper about 12 inches (30.5 cm) square and place on a work surface. Divide the dough into four balls. Place one ball on one parchment sheet and form into a circle. Cover with the other sheet. Using a rolling pin, roll the dough into a thin sheet, about ¼ inch (6 mm) thick. Remove the top parchment sheet.

Line a baking sheet with parchment paper. Using a round 1¾-inch (4.5-cm) cookie cutter (or a shot glass!), cut circles into the thin sheet of dough, using as much of the dough as possible. Remove the excess dough and add it to one of the remaining balls. Peel up a circle of dough and spread with at least 1 teaspoon of filling, then top with another circle. Place on the baking sheet. Repeat with the remaining dough and filling.

Freeze the cookies for at least 3 hours. Serve cold. Store in a tightly sealed container in the freezer for several weeks.

BRAZIL NUTS

Brazil nuts, rich in thyroid-supporting selenium and blood-pressure-controlling minerals, grow on trees in South America's Amazon rainforest. Housed in giant pods containing around 15 nuts per shell, the nuts (actually seeds) grow best in their intact natural ecosystem. One reason is that Brazil nuts can be pollinated only by the jumbo orchid bee, which lives only in the rainforest. So, as a crop, the nut creates an economic incentive for local people to maintain the forest rather than cut it down for farming. And since Brazil nuts are the fruit of the towering Bertholletia excelsa tree, which can live more than 500 years, they mop up a great deal of CO_2.

43 million Americans regularly choose

MEAT-FREE OPTIONS

■ ■ ■

And 86% AREN'T VEGETARIAN or VEGAN

Contributors

TIA BLANCO
FOR BEYOND MEAT

Beyond Meat is on a mission to create meat from plants in an effort to positively impact human health, the environment, constraints of natural resources, and animal welfare.
beyondmeat.com/recipes

DEVORAH BOWEN

KARINA CINNANTE

After twenty years as a vegetarian, Devorah's pursuit of a healthy lifestyle led her to go completely vegan for the past fourteen years. Always experimenting with new recipes, she cooks for family and friends and has been hired to cater intimate dinners.
@theyummyvegan
theyummyvegan.com

PEGGY CHAN

Peggy is the chef and owner of Grassroots Pantry, Hong Kong's leading plant-based restaurant. She's challenging restaurants and chefs to incorporate more plant-based dishes into their menus and to rethink food waste and environmental issues affecting the industry, through a series of pop-up dinners she hosts in collaboration with chefs around the world, called the Collective's Table.
@chefpeggychan
@grassrootspantry

DR. ALKA CHANDNA

PETA

Vice president of laboratory investigations cases at People for the Ethical Treatment of Animals, Alka has been teaching people kindness, compassion, and respect for all living beings for more than twenty-five years.
peta.org
facebook.com/alka1963

JENNÉ CLAIBORNE

SIDNEY BENSIMON

A chef, blogger, and author known for creating healthy and easy-to-make soulful vegan recipes, Jenné became vegan for ethical and animal welfare reasons. Her popular Sweet Potato Soul franchise includes a cookbook, YouTube channel, and website.
@sweetpotatosoul
sweetpotatosoul.com

FRAN COSTIGAN

ROUXBE.COM

Fran Costigan, chef, cookbook author, and director of the Essential Vegan Desserts Course at the Rouxbe cooking school, is internationally renowned as one of the authorities on vegan desserts. A member of Les Dames d'Escoffier, Philadelphia, find Fran exploring Philly's arts, music, and food scenes, and sharing her passion teaching at schools and cooking studios, corporations, cruise ships, and spas, too.
@goodcakesfran
franconstigan.com

GARY DE MATTEI

LINDA LONG

Gary is a vegan activist, chef, filmmaker, actor, producer, director, and writer. A native Californian, he moved to New York in 2008 to be with his life partner, Caryn Hartglass. Together they founded the nonprofit Responsible Eating and Living (REAL).
responsibleeatingandliving.com

JUSTINE DROSDOVECH

MITCH CHAPMAN

An influential vegan health and wellness blogger, Justine is on a mission to inspire others to lead a more compassionate life toward others and themselves, hoping to create a kinder planet for all beings.
thewestcoastvegan.org
@thewestcoastvegan

JANE ESSELSTYN

KEVIN LANE

A science, outdoor, and health educator for more than twenty-five years, Jane is the coauthor of *The Prevent and Reverse Heart Disease Cookbook* and *The Engine 2 Cookbook*.
@janeesselstyn

RIP ESSELSTYN

ENGINE 2 FOR LIFE

A former triathlete and firefighter, Rip is the author of three bestselling books: *The Engine 2 Diet*, *Plant-Strong*, and *The Engine 2 Seven-Day Rescue Diet*.
@engine2diet
engine2diet.com

FORA FOODS

Fora Food's dairy-free butter tastes, bakes, melts, and browns just like dairy butter. Made with upcycled aquafaba blended with a handful of other clean and sustainable ingredients, it is a one-to-one replacer in any recipe ranging from croissants to hollandaise.
@forafoods
forafoods.com

FRY'S FAMILY FOODS

A family-owned business that over the past twenty-seven years has expanded its range of plant-based products to more than thirty-five, Fry's initiated the Meat-Free Mondays movement in South Africa and Australia to educate the public about the benefits of a plant-based diet.
@frysfamily
fryfamilyfood.com

SHARON GANNON

GUZMAN

Founder of the Jivamukti Yoga Method, author of *Yoga and Vegetarianism* and *Simple Recipes for Joy*.
@jivamuktinyc
javamuktiyoga.com

CIERRA DE GRUYTER/ NEXT LEVEL BURGER

CHARLOTTE DUPONT

Cierra is the cofounder and culinary visionary behind Next Level Burger, America's first 100 percent plant-based burger joint, which is all about nourishing people and the planet with non-GMO, all organic veggies.
@nextlevelburger
nextlevelburger.com

CARYN HARTGLASS

CARYN HARTGLASS

Caryn is a vegan activist, chef, singer, host of the podcast *It's All About Food*, and a contributing author to *25 Women Who Survived Cancer*. She cofounded Responsible Eating and Living (REAL) with life partner Gary De Mattei.
@carynhartglass
responsibleeatingandliving.com

THE JACKFRUIT COMPANY

Founded by social entrepreneur Annie Ryu, Jackfruit works with more than 1,000 farming families in southern India to source organically grown jackfruit for the U.S. market.
@thejackfruitco
thejackfruitcompany.com

JUST

Just is on a mission to build a food system where everyone eats well. Just's world-class team of scientists, researchers, and Michelin-starred chefs leverages a one-of-a-kind discovery platform to create delicious, accessible, healthy, and sustainable meat and plant-based food products.
@justforall
justforall.com

SARA KIDD

MELISSA KATHERINE

A former distributor of animal rights and environmental documentaries—including the acclaimed *Kangaroo: A Love-Hate Story*—Sara is a renowned vegan baker and cake designer. Her vegan baking show launched in early 2019.
@vegancakesbysarakidd
sarakidd.com/bake-vegan-stuff-sara-kidd

MARIA KOUTSOGIANNIS

BRE HAFICHUK

Maria, the founder of FoodByMaria, is a Greek Canadian food stylist, fitness enthusiast, and clean-eating queen. Powered by her recovery from an eating disorder, Maria offers her readers approachable plant-based recipes, wellness inspiration, and fitness tips.
@foodbymaria
foodbymaria.com

LENA KSANTI

ANTON ZHELTIKOV

A social media influencer, plant-based chef, educator, and author specializing in healthy-living and raw plant-based cuisine, Lena is the founder of Pure-VeganFood, a plant-based whole foods community.
@pureveganfood
mypureveganfood.com

EMILY LAVIERI-SCULL

EMILY LAVIERI-SCULL

A Brooklyn-based vegan blogger and social media influencer, Emily is passionate about environmental issues, animal rights, and food. She can often be found in the kitchen or hanging out with cats, chickens, or other animals.
@capersandkindness

ALLISON MCLAUGHLIN

ALLISON MCLAUGHLIN

Allison, a photographer and content creator based in New York City, likes plants, a well-organized calendar, thrift stores, things made out of wood, and crystals. You can find her at the farmers' market with way too many mason jars.
@allisonoliviamoon
allisonoliviamoon.com

DIANA MENDOZA

PETA

Diana is PETA's L.A. Companion Animal Program Manager and runs the Let's Fix L.A. program. She represents PETA at city government meetings and assists their Cruelty Investigations Department with local cases.
peta.org

MILKADAMIA

milkadamia

Milkadamia sprouted from the kitchen of a family farm in Australia, near where the macadamia tree originated. The company makes plant-based milks and creamers.
@milkadamia
milkadamia.com

TAAVI MOORE

ABI STEMPAK

A food-obsessed student and blogger from Seattle, Washington, Taavi is passionate about cooking healthy and nutritious vegan meals that are not only easy to make but also provide some fun in the kitchen.
@healthienut
healthienut.com

VICTORIA MORAN

BEN STROTHMANN
PHOTOGRAPHY

A two-time guest on *The Oprah Winfrey Show* and named PETA's Sexiest Vegan Over 50, Victoria is the author of *Main Street Vegan* and twelve other books, host of the *Main Street Vegan* podcast, and director of Main Street Vegan Academy.
@mainstreetvegan
mainstreetvegan.net

JESSICA MURNANE

TAYLOR JARVIS

The creator of the *One Part Plant* movement, host of the *One Part Podcast*, and wellness advocate, Jessica has a certification in plant-based nutrition from the T. Colin Campbell Center for Nutrition Studies.
@jessicamurnane
jessicamurnane.com

JULIA MURRAY

KEZIA NATHE

Julia is a vegan blogger, registered holistic nutritionist, and certified plant-based and raw-food chef. Also an Olympian, a cereal company owner, and a vegan lifestyle coach and educator, she was named one of Canada's Top Vegan Athletes.
@hookedonplants
hookedonplants.ca

NO EVIL FOODS

The makers of artisan plant meat, No Evil Foods is hell-bent on disrupting the food system. Their small-batch plant-based meats challenge the status quo by using simple, sustainable ingredients to re-create a traditional protein experience.
@noevilfoods
noevilfoods.com

NUTPODS

Nutpods are deliciously creamy and made from coconuts and almonds, with zero grams of sugar. This dairy-free creamer is a plant-based alternative to half-and-half. One of the fastest-growing plant-based brands in the U.S., Nutpods is an integral part of wellness programs such as Whole30.
@nutpods
nutpods.com

KIMBERLY PARSONS

The author of *The Yoga Kitchen*, Kimberly is a speaker, lecturer, naturopath, chef, and entrepreneur. She cooks at health retreats all over the world and resides in the English countryside, where she cooks healthy meals for celebrity and private clients.
@_kimberlyparsons_
kimberly-parsons.com

LAURA EDWARDS

PLANTPURE/KIM CAMPBELL

Kim is the author of *PlantPure Nation* and *PlantPure Kitchen* cookbooks, for which she developed more than 250 recipes using no processed oils.
@plantpurenation
plantpurenation.com

REBBL

REBBL is the maker of deliciously nourishing coconut-milk elixirs, soulfully crafted with functional super herbs and organic, ethically sourced ingredients. REBBL donates 2.5 percent of net revenue from the sale of every bottle with a goal to support a future without human trafficking.
@rebbl
rebbl.co

RIPPLE FOODS

Ripple Foods is dairy-free as it should be, creating products that are nutritious and delicious. The company believes that good food should be simple, and that for food to be good, it should nourish your body and leave a small footprint on the planet.
@ripplefoods
ripplefoods.com

ROBIN ROBERTSON

Robin has worked with food for thirty-plus years as a restaurant chef, cooking teacher, and author of more than twenty-five cookbooks, including the best-selling *Vegan Plant, 1,000 Vegan Recipes, Veganize It!, Vegan on the Cheap*, and *One-Dish Vegan.*
@veganwithoutborders
robinrobertson.com

ROBIN ROBERTSON

KLAYTON RUTHERFORD

Klayton grew up in Arkansas and went vegan while working toward an under-graduate degree in English at Arkansas Tech University. He now resides in Texas and works as a membership communications staff writer for the PETA Foundation.
peta.org

FAITH ROBINSON

CHAD SARNO

Chad is the coauthor of *The Wicked Healthy Cookbook* and the VP of Culinary at Good Catch Foods. Chad spent several years at Whole Foods Market as Senior Culinary Educator, Media Spokesperson, and R&D Chef for the Global Healthy Eating program. He is also a contributing author to more than ten cookbooks.
@chadsarno
@wickedhealthy

DAN JONES PHOTOGRAPHY

DEREK SARNO

Derek is the coauthor of *The Wicked Healthy Cookbook* and director of plant-based innovation at UK retail giant Tesco. Formerly senior global executive chef for Whole Foods Market, he was also a chef at a Tibetan Buddhist monastery.
@dereksarno
@wickedhealthy

DAN JONES PHOTOGRAPHY

MIYOKO SCHINNER

MIYOKO'S

Miyoko sparked the rise of the multimillion-dollar plant-based cheese revolution by pioneering a proprietary technology for making dairy from plants. She has redefined the categories of "cheese" and "butter" with complex flavors that have garnered multiple awards.
@miyokoscreamery
miyokos.com

JEFF STANFORD

Jeff is the co-owner of the celebrated Stanford Inn, a compassionate resort on California's Mendocino Coast. At the Inn, he opened Ravens Restaurant, forming a co-creative kitchen where everyone from prep cook to chef contributes to dishes.
@stanfordinnecoresort
stanfordinn.com

SWEET EARTH

Sweet Earth believes that good nutrition, great taste, and sustainable foods are perfect partners, and they've been proving it since 2012 with their globally inspired frozen meals.
@sweetearthfoods
sweetearthfoods.com

DARSHANA THACKER

KATHRYN GAMBLE

Chef and culinary project manager for Forks Over Knives and a graduate of the Natural Gourmet Institute, Darshana was the recipe author of *Forks Over Knives Family* and a lead recipe contributor for *The Forks Over Knives Plan*.
@darshanas_kitchen
darshanathacker.com

TOFURKY

Tofurky thinks food should be easy to prepare, taste good, and also be good for our bodies, animals, and the environment. The company aspires to create foods that are all these things.
@the_tofurky_company
tofurky.com

SAM TURNBULL

SAM TURNBULL

Sam is the creator of the popular recipe blog *It Doesn't Taste Like Chicken* and the YouTube channel by the same name. She is also the author of the rave-reviewed cookbook *Fuss-Free Vegan*.
@itdoesnttastelikechicken
itdoesnttastelikechicken.com

VEGGIE GRILL

Veggie Grill puts a focus on fresh, seasonal ingredients and innovative flavors and has disrupted the fast-casual restaurant landscape by creating craveable, chef-inspired food that is 100 percent plant-based.
@veggiegrill
veggiegrill.com

AMY WEBSTER

RANDY WEBSTER

Amy Webster's passion for protecting farm animals and love of vegan cooking inspired her to start Rainy Day Vegan (RainyDayVegan.com) cooking classes in Seattle. In addition, as a chef with the Humane Society of the United States, she travels around the country training food service professionals on how to make healthy, delicious, plant-based meals.
@rainydayvegan
rainydayvegan.com

About the Authors

NIL ZACHARIAS

Nil Zacharias is the cofounder of Eat For The Planet (eftp.co), 80/20 Plants (8020plants.com), and One Green Planet (onegreenplanet.org). He also hosts the popular podcast *Eat for the Planet with Nil Zacharias* and is a sought-after international speaker based in Los Angeles, California.

GENE STONE

Gene Stone has ghostwritten, cowritten, or written more than 45 books, including such *New York Times* bestsellers as *Forks Over Knives*, *How Not to Die*, *The Awareness*, *The Engine 2 Diet*, and *Living the Farm Sanctuary Life*. He resides in Hudson, New York.

eftp.co

CURIOUS ABOUT A PLANT-BASED DIET AND NEED A LITTLE HELP?

Sign up for 80/20 Plants (8020plants.com)

80/20 Plants is designed to be your best friend for better eating. This service makes it easier than ever to eat a plant-based diet with the help of a simple and effective healthy eating program driven by bite-size daily videos, infographics, and an on-call coach to help you every step of the way. 80/20 Plants is a service that's designed to be Your Best Friend for Better Eating.

Acknowledgments

As always, we'd like to thank all the researchers, scientists, and experts who have studied the impact of our industrialized food system on the environment. You are saving the world, one research project at a time. We are also extremely grateful to all those generous plant-based thinkers, writers, chefs, bloggers, and others who contributed to the making of this book. The world wouldn't be the same without you! Thanks also go out to the following kind people who helped us write this book: Nick Bromley, Kate Good, Chris Hays, Robin Robertson, Miranda Spencer, and Hannah Williams. And, of course, shout-outs to Goji, Toby, Gus, and Julia, our four-legged colleagues.

Finally, we want to thank our excellent agent, Peter McGuigan, as well as the team at Abrams that has been so supportive of our message, especially our gifted editor, Laura Dozier, as well as John Gall, Danielle Youngsmith, Mike Kaserkie, Connor Leonard, Kim Sheu, Jennifer Bastien, and Mamie VanLangen.

Endnotes

1. Yinon M. Bar-On, Rob Phillips, and Ron Milo, "The Biomass Distribution of Earth," *Proceedings of the National Academy of Sciences* (*PNAS*) 115, no. 25 (June 2018): 6506–11, https://doi.org/10.1073/pnas.1711842115.

2. Danielle Nierenberg, "Chapter 2: Rethinking the Global Meat Industry," Worldwatch Institute, www.worldwatch.org/node/3993.

3. European Commission, Joint Research Centre, "Urbanization: 95% of the World's Population Lives On 10% of the Land," ScienceDaily, December 19, 2008, www.sciencedaily.com/releases/2008/12/081217192745.htm.

4. "Livestock and Landscapes," Food and Agriculture Organization of the United Nations (FAO), 2012, www.fao.org/docrep/018/ar591e/ar591e.pdf.

5. http://iopscience.iop.org/article/10.1088/1748-9326/10/12/125012/meta

6. J. B. Veiga et al., "Cattle Ranching in the Amazon Rainforest" (paper presented at the XII World Forestry Congress, Quebec City, Canada, September 2003), www.fao.org/docrep/ARTICLE/WFC/XII/0568-B1.HTM.

7. Tom Aldridge and Herb Schlubach, "Water Requirements for Food Production," *Soil and Water*, no. 38 (Fall 1978), University of California Cooperative Extension, 13017; Paul R. Ehrlich and Anne H. Ehrlich, *Population, Resources, Environment: Issues in Human Ecology* (San Francisco: Freeman, 1972), 75–76.

8. World Wildlife Fund, "A Warning Sign from Our Planet: Nature Needs Life Support," October 30, 2018, www.wwf.org.uk/updates/living-planet-report-2018; Susan Silber, William Velton, "Fact Sheet—Rainforest Animals," http://www.ran.org/fact_sheet_rainforest_animals.

9. World Wildlife Fund, "Global Wildlife Populations: 58 Percent Decline, Driven by Food and Energy Demand," ScienceDaily, October 27, 2016, www.sciencedaily.com/releases/2016/10/161027113306.htm.

10. https://www.ipbes.net/news/Media-Release-Global-Assessment

11. https://www.nationalgeographic.com/foodfeatures/feeding-9-billion

12. World Health Organization, "Drinking-Water," February 7, 2018, www.who.int/news-room/fact-sheets/detail/drinking-water.

13. R. Dennis Olson, "Below-Cost Feed Crops: An Indirect Subsidy for Industrial Animal Factories," Institute for Agriculture and Trade Policy, June 2006, www.iatp.org/sites/default/files/258_2_88122_0.pdf.

14. Betty Hallock, "To Make a Burger, First You Need 660 Gallons of Water," *Los Angeles Times*, January 27, 2014, www.latimes.com/food/dailydish/la-dd-gallons-of-water-to-make-a-burger-20140124-story.html.

15. FoodPrint, "How Industrial Agriculture Affects Our Water," www.gracelinks.org/1361/the-water-footprint-of-food.

16. M. M. Mekonnen and A. Y. Hoekstra, "The Green, Blue and Grey Water Footprint of Farm Animals and Animal Products," Value of Water Research Report Series, no. 48 (December 2010), UNESCO-IHE Institute for Water Education, waterfootprint.org/media/downloads/Report-48-WaterFootprint-AnimalProducts-Vol1.pdf.

17. https://www.treehugger.com/green-food/from-lettuce-to-beef-whats-the-water-footprint-of-your-food.html

18. Alex Park and Julia Lurie, "It Takes HOW Much Water to Make Greek Yogurt?!" *Mother Jones*, March 10, 2017, www.motherjones.com/environment/2014/03/california-water-suck.

19. Vegan Calculator, http://thevegancalculator.com.

20. "Water for Life Decade: Water Scarcity," United Nations Department of Economic and Social Affairs, www.un.org/waterforlifedecade/scarcity.shtml.

21. Peter M. Cox, Chris Huntingford, and Mark S. Williamson, "Emergent Constraint on Equilibrium Climate Sensitivity from Global Temperature Variability," *Nature* 553, no. 7688 (January 2018): 319–22, www.nature.com/articles/nature25450.

22. "Major Cuts of Greenhouse Gas Emissions from Livestock within Reach," FAO, September 26, 2013, www.fao.org/news/story/en/item/197608/icode.

23. Federation of American Societies for Experimental Biology, "Quantifying the Environmental Benefits of Skipping the Meat," ScienceDaily, April 4, 2016, www.sciencedaily.com/releases/2016/04/160404170427.htm.

24 American Journal of Clinical Nutrition, https://academic.oup.com/ajcn/article/78/3/660S/4690010

25 "Overview of Greenhouse Gases," United States Environmental Protection Agency, www.epa.gov/ghgemissions/overview-greenhouse-gases.

26 "Livestock a Major Threat to Environment," FAO, November 29, 2006, www.fao.org/NEWSROOM/en/news/2006/1000448/index.html.

27 Bojana Bajželj et al., "Importance of Food-Demand Management for Climate Mitigation," *Nature Climate Change* 4 (August 2014): 924–29, https://doi.org/10.1038/nclimate2353.

28 Marco Springmann et al., "Analysis and Valuation of the Health and Climate Change Cobenefits of Dietary Change," *PNAS* 113, no. 15 (April 2016): 4146–51, https://doi.org/10.1073/pnas.1523119113.

29 Lisa Pataczek et al., "Beans with Benefits—The Role of Mungbean (Vigna Radiata) in a Changing Environment," *American Journal of Plant Sciences* 9, no. 7 (June 2018): 1577–1600, http://doi.org/10.4236/ajps.2018.97115; Rocío Abín et al., "Environmental Assessment of Intensive Egg Production: A Spanish Case Study," *Journal of Cleaner Production* 179 (April 2018): 160–68, https://doi.org/10.1016/j.jclepro.2018.01.067; M. de Vries and I. J. M. DeBoer, "Comparing Environmental Impacts for Livestock Products: A Review of Life Cycle Assessments," *Livestock Science* 128, nos. 1–3 (March 2010): 1–11, https://doi.org/10.1016/j.livsci.2009.11.007.

30 John Noble Wilford, "Figs Believed to Be First Cultivated Fruit," *New York Times*, June 1, 2006, www.nytimes.com/2006/06/01/science/01cnd-fig.html; Ben Crair, "Love the Fig," *The New Yorker*, August 10, 2016, www.newyorker.com/tech/annals-of-technology/love-the-fig; Editors of Encyclopedia Britannica, "Fig Wasp," Encyclopedia Britannica, updated August 1, 2013, www.britannica.com/animal/fig-wasp.

31 North American Millers' Association; "Study Shows Using Oat Hulls for Power Has Considerable Benefits to the Environment and Human Health," Phys.org, September 25, 2015, https://phys.org/news/2015-09-oat-hulls-power-considerable-benefits.html; "Oats," Sustainable Agriculture Research & Education, https://www.sare.org/Learning-Center/Books/Managing-Cover-Crops-Profitably-3rd-Edition/Text-Version/Nonlegume-Cover-Crops/Oats; Ibrahim Al-Naiema et al., "Impacts of Co-Firing Biomass on Emissions of Particulate Matter to the Atmosphere," *Fuel* 162 (December 2015): 111–20, https://doi.org/10.1016/j.fuel.2015.08.054.

32 Bees/bee-friendly plants/pollination: FoodTank, Environment America, Bee Friendly Food Alliance, New Mexico State University, Encyclopedia Britannica, USFS. Sunflowers: National Sunflower Association; "No Bees, No Food," Environment America, https://environmentamerica.org/programs/ame/no-bees-no-food; "Why Bees Matter," Food and Agriculture Organization of the United Nations, http://www.fao.org/3/i9527en/I9527EN.PDF.

33 James Hamblin, "If Everyone Ate Beans Instead of Beef," *The Atlantic*, August 2, 2017, www.theatlantic.com/health/archive/2017/08/if-everyone-ate-beans-instead-of-beef/535536; Helen Harwatt et al., "Substituting Beans for Beef as a Contribution toward U.S. Climate Change Targets," *Climatic Change* 143, nos. 1–2 (July 2017): 261–70, https://doi.org/10.1007/s10584-017-1969-1; "Pulses and Climate Change," FAO, December 20, 2016, www.fao.org/3/a-c0374e.pdf; "Pulses and Biodiversity," FAO, March 2, 2017, www.fao.org/3/a-i6757e.pdf.

34 Michael Greger and Gene Stone, *How Not to Die: Discover the Foods Scientifically Proven to Prevent and Reverse Disease* (New York: Flatiron Books, 2015); Michael D. Orzolek et al., "Broccoli Production," Penn State Extension, updated June 20, 2005, https://extension.psu.edu/broccoli-production; Jed W. Fahey, Yuesheng Zhang, and Paul Talalay, "Broccoli Sprouts: An Exceptionally Rich Source of Inducers of Enzymes that Protect against Chemical Carcinogens," *PNAS* 94, no. 19 (September 1997): 10367–72, http://doi.org/10.1073/pnas.94.19.10367; M. M. Mekonnen and A. Y. Hoekstra, "The Green, Blue and Grey Water Footprint of Crops and Derived Crop Products," *Hydrology and Earth System Sciences* 15, no. 5 (May 2011): 1577–1600, https://doi.org/10.5194/hess-15-1577-2011; Patricia A. Egner et al., "Rapid and Sustainable Detoxication of Airborne Pollutants by Broccoli Sprout Beverage," *Cancer Prevention Research* 7, no. 8 (August 2014): 813–23, http://doi.org/10.1158/1940-6207.CAPR-14-0103.

35 AusVeg; Potato Council (UK); Tim Hess et al., "The Impact of Changing Food Choices on the Blue Water Scarcity Footprint and Greenhouse Gas Emissions of the British Diet: The Example of Potato, Pasta and Rice," *Journal of Cleaner Production* 112, no. 5 (January 2016): 4558–68, https://doi.org/10.1016/j.jclepro.2015.08.098.

36 Ultimate Citrus, "Florida Citrus and the Environment"; Mekonnen and Hoekstra, "Water Footprint of Crops"; Melanie Fritz, Ursula Rickert, and Gerhard Schiefer, "System Dynamics and Innovation in Food Networks 2010," Proceedings of the 4th International European

Forum on System Dynamics and Innovation in Food Networks, February 8–12, 2010, Innsbruck-Igls, Austria.

[37] "Green Peas," World's Healthiest Foods, www.whfoods.com/genpage.php?tname=foodspice&dbid=55; Paul Alfrey, "Nitrogen Fixing Plants and Microbes," *Permaculture*, July 25, 2014, www.permaculture.co.uk/articles/nitrogen-fixing-plants-microbes; "Nitrogen and Water," U.S. Geological Survey Water Science School, updated January 17, 2017, https://water.usgs.gov/edu/nitrogen.html.

[38] "Eat Smart," Meat Eater's Guide, https://www.ewg.org/meateatersguide/eat-smart/; Katherine Boehrer, "This is How Much Water it Takes to Make Your Favorite Foods," HuffPost, October 13, 2014, https://www.huffingtonpost.com/2014/10/13/food-water-footprint_n_5952862.html; "Pulses and Climate Change"; "Pulses and Biodiversity."

[39] Mekonnen and Hoekstra, "Water Footprint of Crops"; François-Xavier Branthôme, "Prefer Project: The Environmental Impact of Tomato Products," Tomato News, September 19, 2016, www.tomatonews.com/en/prefer-project-the-environmental-impact-of-tomato-products_2_225.html; Kara Pydynkowski et al., "A Life Cycle Analysis for Tomatoes in NH" (research paper, Dartmouth College, 2008), www.dartmouth.edu/~cushman/courses/engs171/Tomatoes.pdf; MindBodyGreen; SimplyHydro.

[40] Hemp Basics, www.hempbasics.com; Jason Daley, "Hemp Makes a Return to George Washington's Farm," *Smithsonian Magazine*, August 24, 2018, www.smithsonianmag.com/smart-news/hemp-makes-return-george-washingtons-farm-180970131; Andrew Leonard, "Can Hemp Clean Up the Earth?" *Rolling Stone*, June 11, 2018, www.rollingstone.com/politics/politics-features/can-hemp-clean-up-the-earth-629589; Rafiq Ahmad et al., "Phytoremediation Potential of Hemp (Cannabis Sativa L.): Identification and Characterization of Heavy Metals Responsive Genes," *CLEAN – Soil Air Water* 44, no. 2 (February 2016): 195–201, https://doi.org/10.1002/clen.201500117.

Illustration sources:

Page 9: Mintel

Page 12: *Eat for the Planet* book

Page 13: World Wildlife Fund

Page 14: *Eat for the Planet* book

Page 15: Journal *Nature*

Page 22: Climatic Change

Page 25: *Eat for the Planet* book

Page 43: *The Lancet Planetary Health*

Page 57: University of Michigan and Tulane University

Page 87: *International Dairy Journal*

Page 103: Federation of American Societies for Experimental Biology

Page 117: Journal *Science*

Page 147: *Eat for the Planet* book

Page 175: The NPD Group

Index

Editor: Laura Dozier
Designer: Danielle Youngsmith
Production Manager: Michael Kaserkie

Library of Congress Control Number: 2018958270

ISBN: 978-1-4197-3441-0
eISBN: 978-1-68335-655-4

ABRAMS The Art of Books
195 Broadway, New York, NY 10007
abramsbooks.com